Urban Geology

Peter Huggenberger • Jannis Epting
Editors

Urban Geology

Process-Oriented Concepts for Adaptive and Integrated Resource Management

Springer

Editors
Professor Dr. Peter Huggenberger
Dr. Jannis Epting
University of Basel
Department of Geosciences
Geological Institute
Applied and Environmental Sciences
Bernoullistrasse 32
4056 Basel, Switzerland
peter.huggenberger@unibas.ch
jannis.epting@unibas.ch

ISBN 978-3-0348-0184-3 e-ISBN 978-3-0348-0185-0
DOI 10.1007/978-3-0348-0185-0
Springer Heidelberg Dordrecht London New York

Library of Congress Control Number: 2011936519

Cover illustrations: Top: Photograph of an excavation pit in the Novartis Campus area. Bottom left: Base of the unconsolidated rock of the Basel area (cf. Chapter 4.1). Bottom right: Groundwater head and temperature development observed in a riverine groundwater monitoring well (cf. Chapter 5.5)

Printed on acid-free paper

Springer Basel AG is part of Springer Science+Business Media (www.springer.com)

Preface

This book reflects the experience of the authors, working in a multidisciplinary team of specialists and scientists on urban geosciences including geology, hydrogeology, hydrogeophysics, river-ecology, and on research projects at the Basel University. Besides the academic activities, the Applied and Environmental Geology (AUG) is in charge of the geological survey of the Cantons of Basel-Stadt and Basel-Landschaft.

Modern quantitative earth-sciences can contribute significantly to finding solutions concerning the sustainable use or subsurface resources in urban environments. The approaches we present in this book are mainly problem oriented. This includes the cooperation of specialists from several universities and institutions with different backgrounds worldwide to find solutions to specific problems related to urban environmental questions.

Urban subsurface resources and especially urban groundwater bodies are particularly vulnerable to environmental impacts, and their rational management is of major importance. Therefore, the development of optimization strategies is necessary. Such strategies should consider simultaneously the numerous impacts on urban subsurface resources, such as infrastructure development or groundwater and geothermal subsurface use.

Often, infrastructure development in urban environments and associated alterations in land use only consider the benefits for the improved infrastructure itself and planning largely takes the pragmatic form of engineering for short-term economic objectives. This often leads to adverse effects concerning quantitative and qualitative issues of subsurface resources including groundwater flow regimes, induced natural hazards, and use conflicts in general.

Although legal frameworks for protection of natural resources have continuously been adjusted in the last decades, damages still occur. Until now, the impacts on natural resources were mostly regarded as solitary limited impacts and examinations of the interactions between them, and other aspects such as possible interactions at a regional scale were not attempted. There are several reasons for this. More attention is paid to purely technological aspects concerning resource management during construction rather than to issues dealing with sustainable resource use as part of our ecosystems. In addition, some projects undertaken under outdated legal frameworks, i.e., some 30 years ago or even longer, would not be approved today because more restrictive laws pertaining to resource use, as well as changed perceptions and policy, now apply.

Currently, our knowledge on subsurface processes is incomplete as concepts for the sustainable use of the urban subsurface are rare. The present legislations and related regulations are confronted with many contradictions which would require a harmonization. These harmonization processes turn out to be very difficult. A discussion on future goals for quantitative and qualitative issues of subsurface resource has just begun. Such present initiatives also include future demands on subsurface resources.

In order to develop strategies for the sustainable use of subsurface resources in urban areas, environmental impact assessments have not only to incorporate aboveground vitiations like noise exposure and air pollution, but also to address the negative impacts on subsurface resources including groundwater flow regimes.

This book presents some alternative approaches for the implementation of adaptive management. Adaptive management schemes of environmental systems start with the definition of particular profiles of systems (i.e., water supply). Together with the identification of system profiles, specific targets are defined that lead to overall goals for particular subsurface resources, in the case of groundwater, i.e., the desired long-term development of urban groundwater resources. As the individual targets may interfere with each other and together not necessarily lead to the desired overall goal, techniques that facilitate the comparison of interference must be applied. This can be accomplished by the development of scenarios and the implementation of equivalence and acceptance criteria.

The conceptual approach we propose includes the combination of instruments that allow to adequately identifying influences of the various single impacts on the complete environmental system. Both impacts that only affect the system in its immediate vicinity and impacts with influence on the system on a regional scale are considered. There are four main elements which are important for a successful management of urban subsurface resources: (1) Efficient management of subsurface data and data mining to provide geological data in 3D; data should be organized in such a way that fast data access is provided; (2) Specific field investigations and experiments to study the relevant processes in urban environments and to provide adequate boundaries for modeling approaches; (3) Development of tools for intelligent analysis of subsurface monitoring data and the setup of geological, hydrogeological, or geotechnical models; and (4) The development and implementation of adaptive management concepts at different scales as a base for the setup of scenario techniques in decision processes. Based on these elements, comparative studies as well as scenario development are focused on predefined development goals.

An important aspect of resource management in urban areas is the availability of geological and hydrological information. Generally, large amounts of data are available that are spread at different institutions. The availability of these data often is difficult and its preparation for specific questions time consuming. This was the main reason to setup a geological database for northwestern Switzerland, consisting of a systematic data collection, an analysis of drill-core data, including the administration of metadata from geological and hydrological reports. The database can be connected to a Geographical Information System (GIS) for 3D structural analysis. Together with further hydrological data, the database represents a unique data source that is suitable for empirical studies and hypothesis testing in the domain of quantitative information fusion of urban geological or hydrological questions.

The book chapters integrate existing and new scientific knowledge, methods, and tools into these new concepts. Such an approach facilitates the implementation of the Water Framework (WFD) and Habitats Directives (HD) as well as a better management of subsurface resources. Main target groups addressed include professional hydrogeologists and geologists, urban planners and water supply engineers, environmental agencies, universities, as well as students in hydrogeology, planning, water supply, and environmental sciences.

The topics illustrated in this book have their origin in projects in the urban region of Basel, northwestern Switzerland. The examples deal with questions which have practical as well as research character. Almost all topics are also relevant for other urban areas and the sustainable use of subsurface resources in general.

Basel, Switzerland

Peter Huggenberger
Jannis Epting

Acknowledgments

The editors thank all contributors to this book for their efforts in collaborating in the various chapters. Special gratitude is expressed to Annette Affolter for her endurance in preparing all illustrations and tables and Eva Vojtech for her critical review. Furthermore, we acknowledge the financial support of the Freiwillige Akademische Gesellschaft (FAG), the Swiss Academy of Sciences (SNAT), and Hoffman LaRoche. Last but not least, we thank Springer for the opportunity of publishing this book.

Contents

Contributors

Internal Authors

Annette Affolter, annette.affolter@unibas.ch

Horst Dresmann, horst.dresmann@unibas.ch

Jannis Epting, jannis.epting@unibas.ch

Peter Huggenberger, peter.huggenberger@unibas.ch

Rebecca M. Page, rebecca.page@unibas.ch

Stefan Scheidler, stefan.scheidler@unibas.ch

Stefan Wiesmeier, stefan.wiesmeier@unibas.ch

Eric Zechner, eric.zechner@unibas.ch

Ali Zidane, ali.zidane@unibas.ch

Applied and Environmental Geology, Geological Institute, Department of Geosciences, University of Basel, Bernoullistrasse 32, 4056 Basel, Switzerland

External Authors

Christoph Butscher Massachusetts Institute of Technology, Massachusetts avenue 77, Cambridge, MA 02139–4307, USA, butscher@MIT.EDU

Donat Fäh ETH Zürich, Schweiz. Erdbebendienst (SED), Sonneggstrasse 5, 8092 Zürich, Switzerland, faeh@sed.ethz.ch

Daniel Gechter Kellerhals + Haefeli AG, Kapellenstrasse 22, 3011 Bern, Switzerland, daniel.gechter@sunrise.ch

Ralph Kirchhofer Fachstelle für Geoinformation, Grundbuch- und Vermessungsamt, Münsterplatz 11, 4001 Basel, Switzerland, ralph.kirchhofer@bs.ch

Markus Konz RMS, Stampfenbachstrasse 85, 8021 Zürich, Switzerland, markus.konz@rms.com

Edi Meier Edi Meier + Partner AG, Geophysik und Geotechnik, Technopark-Winterthur, Jägerstrasse 2, 8406 Winterthur, Switzerland, edi.meier@emp-winterthur.ch

Christian Regli GEOTEST AG, Promenade 15, 7270 Davos Platz, Switzerland, christian.regli@geotest.ch

Douchko Romanov Institute of Geological Sciences, FU Berlin, Malteserstrasse 74-100, 12249 Berlin, Germany, douchko.romanov@fu-berlin.de

Jelena Simovic Rota Cantonal Office for the Environment and Energy, Hochbergerstrasse 158, 4019 Basel, Switzerland, jelena.rota@gmail.com

Chapter 1
Content

Peter Huggenberger and Jannis Epting

The various research topics that are illustrated in this book have their origin in projects in the region of Basel, Northwestern Switzerland. They deal with questions which have practical as well as research character in the domain of "urban geology." In the following, a brief overview of the contents in the various book chapters is given.

1.1 Chapter 2: Settings in Urban Environments

This chapter summarizes the common settings in urban environments, including a general description of the value, functions and characteristics of urban resources (water, energy, materials, etc.) as well as a statement about the challenges for environmental sciences.

The chapter also includes an asset of present protection and management strategies. The necessity to provide decision support for questions arising in the context of "urban geology" as a service for the public domain will be highlighted.

1.2 Chapter 3: Hypotheses and Concepts

This chapter introduces some of the main hypotheses and presents several concepts for adaptive and integrated resource management in urban areas. The methods are discussed together with the basic principles for the sustainable use of urban resources.

As urban environments are continuously changing and settings are spatiotemporal highly heterogeneous, such approaches allow to plan and control sustainable infrastructure development with respect to natural resources. Prerequisites are analyses of resource systems and an inventory of current and future profiles of

P. Huggenberger and J. Epting (eds.), *Urban Geology: Process-Oriented Concepts for Adaptive and Integrated Resource Management*, DOI 10.1007/978-3-0348-0185-0_1,

such systems together with the definition of targets and goals for specific urban regions (Sect. 3.1).

We focus on how to advance an understanding of some of the relevant process in urban environments and on developing methods for testing hypotheses. In a next step, we outline risk profiles for subsurface resources, which comprise the determination of principal hazards or risk patterns for subareas and different resource users. This also includes the identification, localization, and capture of the relevant processes that lead to specific risk situations (i.e., conflicts and hazards from geothermal energy use, diffuse and point source pollution, microbial pollution through river–groundwater interaction, etc.). Thereby, the detection of risk situations is the basis for differentiated subsurface resource protection measures (Sect. 3.1).

The management of resources in urban areas requires a definition of manageable units of the subsurface. The delineation of such units not only is relevant for the exploitation of subsurface resources, but also allows to define boundaries and to derive fluxes of heat and mass including water compounds across these boundaries (Sect. 3.2).

We present a sustainable regional planning concept for the use and protection of water resources that allows us to address both spatial and temporal aspects of groundwater vulnerability. Furthermore, we discuss the role of quality control systems, which include the monitoring of physical, chemical and microbiological parameters, the definition of Critical Control Points (CCPs) as well as flux calculations, which can be derived from groundwater modeling (Sect. 3.3).

In the context of the ongoing debates on the impact of anthropogenic and climate change to quantitative and qualitative aspects of groundwater resources, we evaluated and summarized the present state of the groundwater temperatures in the city Basel. In three parts, we discuss (1) several positive and negative feedback mechanisms concerning water and thermal budgets and the impacts of climate change in urban environments; (2) the effects of predicted climate change on groundwater vulnerability in urban environments; and (3) analyses of historical groundwater temperature data to delineate different zones of urban groundwater bodies (GWB) and to optimize future observation networks (Sect. 3.4).

1.3 Chapter 4: Methods

A unique urban system is presented where geological, hydrogeological and hydrological data are systematically collected, verified and integrated into a comprehensive database and Geographic Information Systems (GIS) and from there into geological and hydrogeological models. This basis of information also allows us to develop tools for seismological prediction of subsurface behavior during major earthquakes. The provision of tailored database and GIS applications, including preliminary data analysis, 2D and 3D data as well as geostatistical analysis will be highlighted. In this chapter, we also address general statements regarding the role of data for urban geological and hydrogeological issues (Sect. 4.1).

In a next step, we present some basic elements for adaptive resource management, which include (1) the setup of adequate observation networks for monitoring; (2) selection of appropriate modeling tools; and (3) the definition and realization of specific field measurements to close existing knowledge gaps. We discuss some general thoughts concerning the optimal design of observation networks and the appropriate selection of measurement parameters. Further, we illustrate the choice of some available geological and hydrogeological modeling approaches for different environmental questions (Sect. 4.2).

As an example for comprehensive field investigations we present some hydrogeophysical research methods, including their applicability in urban environments. We show that the application of such methods allows a spatial continuous characterization of the subsurface and can be used for a nondestructive mapping and monitoring (Sect. 4.3).

Most urban aquifers are characterized by a high natural and anthropogenic heterogeneity of the subsurface as well as a large spatial variability of hydraulic parameters. Therefore, detailed knowledge of subsurface structures is an important prerequisite for the understanding and solution of specific problems related to adaptive resource management. We present some of the existing concepts and methods for the assessment and description of subsurface heterogeneity. Emphasis is placed on structure analyses using geostatistical approaches (Sect. 4.4).

When studying geological and hydrogeological processes a huge amount of spatiotemporal data accumulate, which have to be analyzed and interpreted. In this chapter, we present methods such as nonlinear statistics that allow the extraction of relevant information by hiding unnecessary information of complex datasets (Sect. 4.5).

1.4 Chapter 5: Examples and Case Studies

In this chapter, we illustrate results of case studies from the region of Basel, Northwestern Switzerland.

In a first set of case studies we address protection issues of groundwater production and river restoration in urban areas, with a focus on drinking water supply aspects. We present protection schemes for several major drinking water supplies in the region of Basel. We focus on hydrogeoecological issues in the context of river restoration projects in urban environments. Urbanization in the last century created a series of environmental problems such as flooding, groundwater pollution and ecological changes, including a decrease of characteristic habitats of riverine landscapes together with a drastic reduction of species. With three examples, we illustrate strategies to integrate hydrogeoecological aspects in an early planning process of engineering projects as drinking water and flood protection measures or river restoration in urban areas. Further we focus on the setup of monitoring networks and modeling tools, river–groundwater interaction, aquifer heterogeneity, and the reconciliation of water engineering measures along rivers.

In a second set of case studies, we address engineering and hydrogeological questions that emerged during the development of urban infrastructure projects in the region of Basel. Here, we focus on groundwater management and protection issues during and after completion of two infrastructure development and upgrading projects.

In a third set of case studies, we encompass management concepts as well as monitoring, modeling and remediation strategies for contaminated sites in trans-boundary settings. In a first case study, we discuss strategies to understand and predict the cumulative effects of the numerous single impacts on groundwater resources during a major suburban development project. In a second case study, we illustrate the development of groundwater pollution in a heavily industrialized groundwater protection area during the last decades.

In the fourth set of case studies, we address karst in urban environments. Groundwater circulation in evaporate-bearing horizons and the resulting evolution of karst frequently causes geotechnical problems such as land-subsidence or collapses. Such processes are of particular concern in urban areas where soluble geological formations coincide with vulnerable infrastructures as transportation systems. In this chapter, we focus on two case studies where subrosion, land-subsidence, and impacts on infrastructures have been observed. The case studies allow the illustration of the complex interrelations between natural phenomena and processes that are induced by present-day engineering and subsurface activities in the region of Basel.

In the fifth set of case studies, we address the use of shallow geothermal energy in urban environments. Increasing geothermal energy use can exceed the subsurface potential for different heating and cooling systems and effect groundwater quality. Currently, in most urban areas, regulations for water resource management and geothermal energy use are sparse and limited to the rule “first come, first served.” In this chapter, we focus on concepts for monitoring and modeling the influence of geothermal systems as well as on the provision of suitability maps for site evaluation. In the first case study, we present a concept that allows to rapidly evaluate proposed drilling sites that are suitable for geothermal use. In the second case, we present a thermal groundwater management concept on the basis of developed monitoring and modeling tools.

In a sixth set of case studies we deal with natural hazards in a regional context, including earthquakes and earthquake risk reduction, major flood events, and flood protection measures.

Chapter 2
Settings in Urban Environments

Peter Huggenberger and Jannis Epting

The history of subsurface resource use in urban areas is generally dominated by the activities during industrialization and even more so since the 1950s. If we want to understand the present condition of the quantitative and qualitative status of subsurface resources, especially concerning water resources in urban areas, we need to know the changes that occurred during this time period. Such changes include infrastructure development as the use of the subsurface for the construction of traffic lines which often interfere directly with water resources.

These changes to the subsurface structure and the numerous anthropogenic impacts make urban geological and hydrogeological issues complex. Additionally, innovative concepts for efficient management and resource protection for the subsurface are sparse. Historically, "low-level" resource management took place over a long time period. At the beginning of the last century, diseases and severe health problems made society aware of the negative impacts of intense and abusive resource exploitation. Especially in urban environments, the variety of pollution is generally more diverse compared to rural areas. This deficit causes severe problems today, when dealing with questions about the use of groundwater, the construction of traffic lines, waste disposal sites, or geothermal use of the subsurface. It also can be expected that these problems will accelerate in the near future.

About 70% of the European population lives in urban areas, which cover in total about 25% of the total European territory (EEA 1999). More than 40% of the water supply of Western and Eastern Europe and the Mediterranean region come from urban aquifers. For optimized and sustainable water resource use in urban regions, therefore, efficient and cost-effective management tools are essential to maintain quality of life and to ensure that water is available for use by future generations (Eiswirth et al. 2003). Sustainable use of soil, groundwater, and other important resources in urban areas is hence a key issue of European environmental policy (Prokop 2003).

Whereas rules for land and surface resource management exist, rules for subsurface planning and management (e.g., "invisibility" of water resources or geothermal energy) are almost absent. Due to the lack of rules for urban subsurface use, current planning procedures do not account for the interactions between different

P. Huggenberger and J. Epting (eds.), *Urban Geology: Process-Oriented Concepts for Adaptive and Integrated Resource Management,* DOI 10.1007/978-3-0348-0185-0_2, usages of the subsurface and consequently subsurface resources use is most likely inefficient and can lead to considerable risks. One example is the observed areal subsidence in the Upper Rhine region (Adlertunnel in Basel-Landschaft, Switzerland). Another example is the observed land uplift as a result of well construction for the geothermal use of the shallow subsurface that came along with the connection of confined aquifers with rocks that are susceptible to swelling (Staufen, southwest Germany).

To develop concepts and methods for sustainable subsurface use in urban areas, environmental impact assessments not only have to include above-ground impairments, such as ground motions with effects on existing buildings and infrastructures, noise exposure and air pollution, but also the negative impacts on subsurface resources. In order to develop rules for the use of urban subsurface space, the complexity of emerging problems has to be broken down into elements. Therefore, the challenge is to integrate innovative concepts into effective, holistic plans for sustainable resource planning and management.

This chapter summarizes the settings in urban environments and highlights how they differ from rural areas. Further we focus on infrastructure development and use conflicts in urban areas, legal backgrounds as well as the general settings of the described case studies.

2.1 Infrastructure Development

Generally open space in urban areas is very rare. Therefore, the subsurface in urban areas is used more frequently for the growth of city infrastructure and traffic lines. Such constructions can temporarily affect urban groundwater systems during the construction period and permanently after completion. Subsurface constructions inevitably increase the pressure on urban groundwater resources and often involve a reduction of cross-sectional groundwater flow and aquifer-storage capacities. As a result subsurface resources are subject to ongoing adaptations under changing hydrological and technical boundary conditions.

Often infrastructure development and associated changes in land-use largely takes the pragmatic form of engineering for short-term benefits. New subsurface infrastructure often is realized under difficult geotechnical and hydrogeological conditions. In particular, tunnel construction in unconsolidated rocks and below the water table can lead to a higher risk of surface subsidence or collapse. To maintain the rapid pace of city life while ensuring that safety standards are met on construction sites, geotechnical measures such as cement injections for subsurface stabilization in unconsolidated rock are commonly used. The potential for hazards during construction is considerably high. Substances used on the construction site as remains of cement injections as well as the used substantives can lead to contamination. Furthermore, such stabilization measures may lead to adverse effects on groundwater flow regimes with regard to quantity and quality of water resources.

In some cases, changes in water fluxes and new created water ways can have negative impacts on adjacent infrastructures.

For this reason, constructions within the groundwater should be limited to the necessary. In case no other solutions are available, the question should be raised on how the impact of such constructions can be minimized. Altogether, constructions of infrastructure facilities (e.g., installation of shallow geothermal systems, subsurface dissolution mining for salt production, power lines, etc.) should take place under controlled hydraulic conditions, including the continuous measurement of hydraulics as well as further physical and chemical (T, EC, pH, Turbidity, etc.) or geotechnical parameters (inclination measurements, etc.).

In Chap. 3, we introduce some concepts for a controlled and sustainable infrastructure development in urban areas and apply them to case studies. The concepts base on the understanding of the principal hydrogeological and geotechnical processes in urban areas.

2.2 Use Conflicts in Urban Areas

Numerous use conflicts have to be considered in urban areas, such as municipal and industrial groundwater use or shallow and deep geothermal energy use. Additionally, historical aspects of the development of urban areas have to be considered (contaminated areas, infrastructure and public transportation development in the shallow unconsolidated and consolidated subsurface, water supply, subrosion processes, etc.). While some usages only temporarily affect urban groundwater systems, e.g., during scheduled operation of water use (day/night, winter/summer) or during the construction of infrastructure, other impacts are permanent, like the reduction of cross-sectional groundwater flow and aquifer-storage capacities (see above).

Competing usages in urban areas further include:

1. The extraction for drinking water supply and industrial processes.
2. Thermal groundwater use, including groundwater extractions and injections for cooling processes and heat production.
3. Water engineering measures, including flood control, construction site drainages, construction parts reaching into the aquifer and storm water management.
4. The growing use of water for modern city architecture like fountains, small streams, ponds, lakes, water-plays, etc.

It is likely that a higher density of the mentioned projects will lead to more use conflicts in the future. These different usages can result in significant changes in groundwater quality and dynamics of both local and regional groundwater flow regimes. It is an important issue of adaptive resource management (Sect. 3.1) to understand the changes due to urban infrastructure development. Further examples or topics of use conflicts are discussed in separate book chapters.

2.3 Legal Background

Although legal frameworks for subsurface protection and policy strategies have continuously been adjusted in the last decades, considerable damages to subsurface resources and groundwater flow regimes still occur. Existing legislation only partially includes the quantitative conservation of subsurface resources as well as a more restricted approval of infrastructures and incorporated construction site drainages.

There are several reasons for this discrepancy:

1. During infrastructure development, more attention is paid to purely technological and constructional problems concerning subsurface resource management rather than to issues dealing with sustainable resource use or possible interferences with historically polluted industrial areas.
2. Some projects undertaken under outdated legal frameworks, i.e., some 30 years ago, would not be approved today because more restrictive laws pertaining to subsurface resources, as well as changed perceptions and policy concerning these resources and its sustainable use, now apply.
3. Subsurface resource protection in urban areas is still focused mainly on documentation of changes in groundwater quality and flow regimes, like maintaining local flow capacities and preventing a significant lowering of the groundwater table. Less attention is paid to the prediction of future demands and to the management of subsurface resources.
4. Until now, the impacts of the various subsurface resource users were only regarded as solitary limited impacts and examinations of the interactions between them were not attempted. Other aspects, such as possible interactions with former industrial sites were often neglected.

With regards to urban aquifer systems, several legal aspects have to be considered. This includes the protection of aquifers which should not be (a) connected in such ways that quantitative or qualitative changes of the groundwater flow regime may occur and (b) essentially and permanently reduced in storage volume and cross-section for flow by constructions into usable aquifers. Often regulations include restrictions for reduced flow through capacities in the order of magnitude of 10%.

2.4 General Settings of the Outlined Case Studies

We present several case studies from the Basel area that illustrate a number of questions related to urban development (Fig. 2.1). The Basel region, which borders both Germany and France, acts as a vital regional as well as interregional traffic junction and represents one of the three designated economic centers of Switzerland. Moreover, Basel has a variety of natural environments, as well as highly vulnerable groundwater systems in river valleys and adjacent karstified areas.

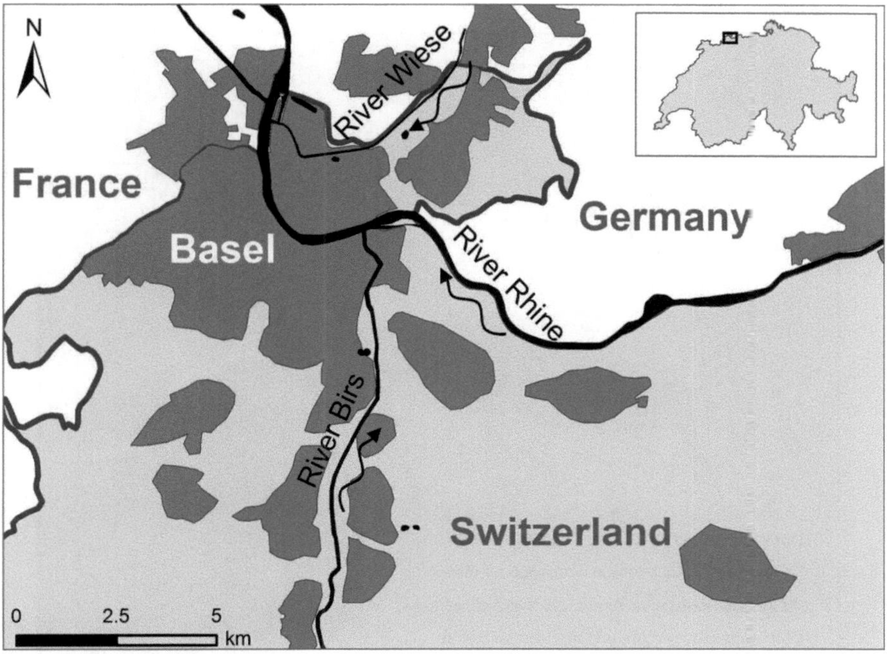

Fig. 2.1 The urban agglomeration of Basel

The subsurface geological composition and structure is summarized in the following box. The existence of evaporites and mixtures of marl-bearing evaporites in the Triassic formations as well as the fact that Basel is located in the seismologically most active area of central Europe comes along with the potential occurrence of geohazards. Some of these hazards are natural; others are triggered by human activities. We use the presented case studies to develop and to test strategies which support a sustainable long-term development of the urban environment and subsurface resources.

Geological Setting of the Basel Area

A general overview of the geology in the Basel area is given in Fig. 2.2, with the stratigraphic units defined in Table 2.1. The dominant tectonic feature is the eastern master fault of the Southern Rhine Graben separating the Rhine Graben and Tabular Jura. The vertical offset at the border fault of the Rhine Graben is about 1,400 m. Within the Rhine Graben (on the down-thrown side), the Mesozoic strata (Triassic to Jurassic; UPM, MES, PCB) are covered by 500–1,000 m of Cenozoic sediments. Three main Graben structures can be distinguished in the Basel area. The Cenocoic sediments in the area were deposited in the asymmetric syncline of St. Jakob-Tüllingen (SJT) adjacent to the main border fault. To the west the Rhine Graben then rises to the Horst of Basel (HB). Further west follows the "Allschwil fault zone" (AF), which sets

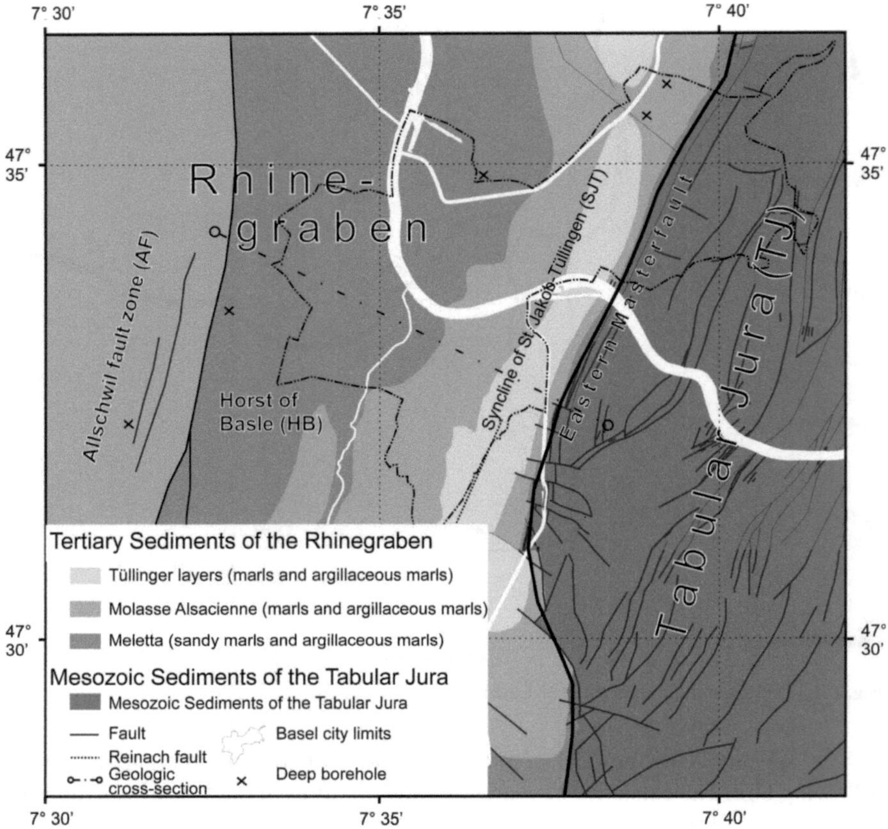

Fig. 2.2 Geological overview of the Basel area (modified after Noack et al. 1999)

Table 2.1 Stratigraphic units represented in the 3D-model and their abbreviations

Abbreviation	Stratigraphy
QUA	Quaternary sediments
TUE	Tüllingen layers (Tertiary); marls and argillaceous marls
ALS	Molasse Alsacienne (Tertiary); sandy marls
MEL	Meletta layers (Tertiary); sandy and argillaceous marls
UPM	Lower Tertiary/first Mesozoic sediments; Sannoisien (Tertiary) and upper Mesozoic sediments down to Lias
MES	Lower Mesozoic; Mesozoic sediments of the Lias and older
PCB	Lowest Mesozoic sediments ("Buntsandstein"), Paleozoic sediments ("Rotliegendes") and crystalline basement

off the Graben sediments in the order of 500 m. The profile in Fig. 2.3 illustrates these structures.

The sedimentary composition of the Cenozoic layers to the west of the Rhine Graben master fault is known by outcrops located predominantly at the Graben

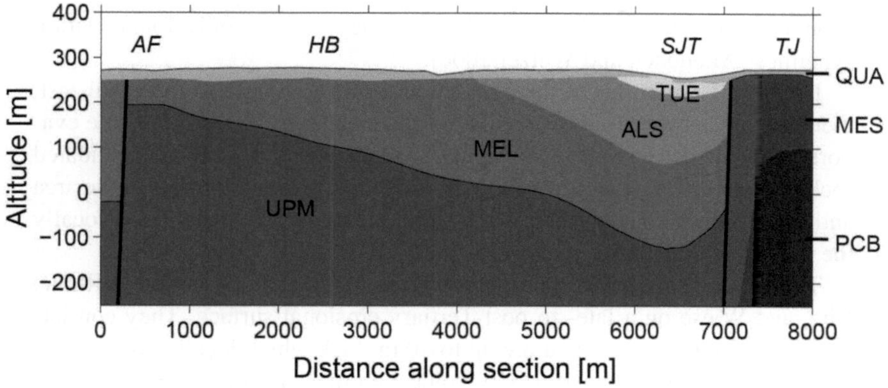

Fig. 2.3 Stratigraphic units below the surficial Quaternary sediments in the Basel area (line of section shown in Fig. 2.2). The *three letter codes* for the stratigraphic units are explained in Table 2.1. The dominant seismic contrast inside the Rhine Graben is between the units MEL and UPM, indicated with a line in the section (Kind 2002)

borders, six deep drill holes (>1,000 m), and a dense network of more than 10,000 boreholes (0 to <1,000 m) drilled, e.g., for geotechnical and groundwater investigation purposes (Sect. 4.1). The following formations can be distinguished in the boreholes (Table 2.1): argillaceous marls and clays of the Meletta layers (MEL, max. 350 m thick), the sandy "Molasse Alsacienne" (ALS, max. 350 m thick) and Tüllinger layers (TUE, max. 200 m thick) which consist of calcareous to argillaceous marls alternating with freshwater carbonates.

To the east of the Rhine Graben master fault the Dinkelberg block as a part of the Tabular Jura is bounded by larger faults such as the Rhine Graben master Fault, the Kandern Fault, the Werratal Fault, and the Zeinigen Fault. The block boundary is poorly defined in the South. The entire Dinkelberg block is characterized by a set of NNE–SSE striking narrow Graben structures. The southern part of the Dinkelberg block, the ESE–WNW striking Adlerhof Anticline, represents a compressive structure.

The basement rocks from the southern Black Forest have been affected by regional metamorphism, large-scale thrust tectonics, and extensive magmatic activity during the Variscan orogeny (e.g., Hann and Sawatzki 2000). At the end of the Variscan orogeny numerous intramontane basins were formed, as for example the so-called Permo-Carboniferous Basin of Northern Switzerland from the Burgundy to the Lake Constance. With the onset of the Triassic transgression, coastal and marine conditions developed. The principal decollement horizons of the Jura Mountains are Middle and Late Triassic evaporites. From the Late Triassic until the end of the Mesozoic, marine conditions prevailed resulting in a sedimentary stack between 1,000 and 1,500 m thick, consisting mainly of limestones, marls, and clays. Cretaceous deposits occur only west of the study area. Tertiary deposits are conserved in the Bresse and Rhine Graben as well as in the Molasse Basin. In the Jura

Mountains, erosional remnants of Tertiary deposits are only found within synclines (Allenbach and Wetzel 2006).

Due to the slight dip of the Triassic and Jurassic strata to the Southeast, there are areas, where subrosion altered the mechanical behavior of the evaporite zone, which at several locations can resemble more to a unconsolidated rock sequence. The Hauptmuschelkalk is one of the main aquifers in the area and often shows characteristic karst phenomena, which also altered locally the mechanical behavior of this unit (Sect. 5.4).

The Quaternary sediments were deposited in the main river valleys Rhine, Birs, and Wiese on a late- to post-Tertiary erosional surface. They consists mainly of fluvial gravels that are up to 40 m thick which locally are covered by Loess. During deglaciation and Holocene times, a series of river terraces formed, separated from each other by terrace bluffs. The geometry of the surface morphology gives some indication on the development of the valley fills. Due to the high permeability and porosities, the fluvial gravel deposits represent the most productive groundwater reservoirs in the area. Pronounced sedimentary structures and textures in the gravel deposits are important for the interpretation of the complex groundwater flow field and flow regime.

The Basel area includes a series of important local and regional scale water supplies, several floodplains in densely populated river valleys, as the floodplains of the Birs River (12 km^2) and the Wiese River (6 km^2) as well as the heavily industrialized area in the floodplain of the Rhine River (64 km^2). All sites represent important groundwater production areas in park-like natural recreation environments, surrounded by urban agglomeration, industry, contaminated sites and traffic. Drinking water supply competes with other interests or demands such as river training, flood control, recreation as well as urbanization and changes of land-use.

In the last 10 years, these sites have been equipped with extensive groundwater monitoring systems. At the same time, high-resolution geological and hydrogeological models were set up and calibrated with long-term datasets that allow comprehensive investigations of subsurface resources, groundwater flow regimes, and the description of relevant boundary fluxes. The models have predictive capabilities and have already been successfully used for scenario development. These already existing tools provide substantial contributions to the understanding of hydrogeological processes and are the basis for hypothesis testing.

References

Allenbach RP, Wetzel A (2006) Spatial patterns of Mesozoic facies relationships and the age of the Rhenish Lineament: a compilation. Int J Earth Sci (Geol Rundsch) 95:803–813. doi:10.1007/s00531-006-071-0

EEA (1999) Environment in the European Union at the turn of the century. European Environment Agency, Copenhagen, Denmark

Eiswirth M, Hötzl H, Cronin A, Morris B, Veselič M, Bufler R, Burn S, Dillon P (2003) Assessing and improving sustainability of urban water resources and systems. RMZ Mater Geoenviron 50:117–120

Hann HP, Sawatzki G (2000) Neue Daten zur Tektonik des Südschwarzwaldes. Jber Mitt Oberrhein Geol Ver 82:363–376

Kind F (2002) Development of microzonation methods: application to Basel, Switzerland, Ph.D. thesis, ETH Zürich, Available in electronic form from http://www.ethbib.ethz.ch

Noack T, Fäh D, Kruspan P (1999) Erdbebenmikrozonierung für den Kanton Basel-Stadt. Geologischer Bericht Nr. 24, Landeshydrologie und -geologie, 83p

Prokop G (2003) Sustainable management of soil and groundwater resources in urban areas. In: Proceedings of the 2nd IMAGETRAIN cluster meeting, Krakow, Poland, 2–4 Oct 2002

Chapter 3
Hypotheses and Concepts

Peter Huggenberger, Jannis Epting, Annette Affolter, Christoph Butscher, Stefan Scheidler, and Jelena Simovic Rota

Within this chapter, we present and discuss several hypotheses and some concepts which we consider as important for urban geology. The first section deals with adaptive subsurface and groundwater resource management in urban areas with a focus on the definition of "system and risk profiles" (Sect. 3.1). The second section discusses the importance and role of "flow across boundaries" (Sect. 3.2). The third section describes an approach for the assessment of "vulnerability" of urban groundwater resources and includes a discussion on how to define "quality control systems" (Sect. 3.3). In the last section, we discuss impacts of anthropogenic and climate change to quantitative and qualitative aspects of groundwater resources in the city Basel (Sect. 3.4).

The taken measures that are addressed in the concepts are directed towards a better understanding of urban subsurface systems in order to improve the base for future decisions. They can be used as an asset framework or tool for subsurface planning in the evaluation of individual projects as well as for optimization of subsurface resource management in urban areas.

The sustainable use of subsurface resources in urban areas requires a profound understanding of the geological and hydrogeological processes and particularly of the short- and long-term effects of human activities today and tomorrow. This includes an adequate evaluation and consideration of variable hydrologic and geotechnical boundary conditions that influence the subsurface and groundwater flow regimes.

Currently, the knowledge of the complex interference between natural and anthropogenic impacts on these processes is incomplete. In the past, issues related to urban geology and hydrogeology were typically taken at the level of individual projects. Groundwater protection strategies in urban areas are mainly oriented to requirements of particular sectors, e.g., protection zones for pumping wells. Quality control systems for groundwater resources often are restricted to measuring a limited set of parameters and defining thresholds on a routine basis. However, it is the sum of all impacts together with their interaction in time and space which explains the present state of our subsurface resources. Since a few years also time series of groundwater data are analyzed and interpreted in the context of the

P. Huggenberger and J. Epting (eds.), *Urban Geology: Process-Oriented Concepts for Adaptive and Integrated Resource Management*, DOI 10.1007/978-3-0348-0185-0_3,

relevant system processes and changing hydrological and operational boundary conditions.

Also existing legal frameworks for environmental protection usually focus on monitoring local data of concentrations and on limiting values of emissions rather than trying to understand the fundamental processes and long-term changes. This practice might be suitable for certain domains (e.g., during short-term impacts), but makes it difficult to implement sustainability concepts in regard to subsurface resources.

Planning concepts for urban development should therefore focus more on the efficient use of subsurface resources, the estimation of the most relevant processes and parameters that are related to possible hazards and environmental degradation, as well as long-term planning and control of resource exploitation. To optimize the regional management of urban subsurface resources, it is necessary to understand these processes and to reach a consensus on long-, medium-, and short-term goals. Such planning concepts include innovative approaches, which take into take account of the complexity of the system and facilitate the adequate quantification of site-specific aspects. This also includes considering the consequences of cumulative effects of individual projects at a larger scale. Such approaches can be summarized as adaptive groundwater management concepts as outlined by Fatta et al. (2002), Eiswirth et al. (2003), Pahl-Wostl et al. (2005), and Epting et al. (2008). However, these concepts have rarely been applied successfully in urban planning. Strategies for a sustainable development of the urban subsurface are at least twofold, including the optimization of (A) new locations for subsurface infrastructure development and groundwater use (optimal spatial integration into existing supply networks and consideration of existing and new subsurface infrastructures) and (B) short- and long-term subsurface and groundwater management programs. Such optimizations allow answering a series of important questions concerning the sustainable use of resources. Some of the specific questions that are addressed in the various book chapters include:

1. Which are the determining processes, including their spatiotemporal scales that have to be considered for urban surface and subsurface development in general but also during the development of particular systems?
2. What is the influence of the increasing density of subsurface constructions and their interactions on hydrological, hydrogeological, geotechnical, and other environmental issues and how can a sustainable use of subsurface resources be achieved despite this pressure?
3. How can urban resources be used in a sustainable way, considering future subsurface infrastructure development as well as geothermal energy use?
4. How can subsurface processes be embedded into strategies and planning procedures for urban development?

Overall, a strong link to process-oriented research at universities could build a bridge to provide a fast transfer of knowledge from the universities to the public services in charge for resource protection.

3.1 System and Risk Profiles

Peter Huggenberger and Jannis Epting

The basic principles of adaptive subsurface and water resource management include the identification of the current profiles of environmental systems and risks at relevant scales. In combination with the formulation of specific targets this can lead to the achievement of long-term development goals. Similar approaches are common in the economic world. However, the experience shows that the discussion on how such approaches can be implemented in the context of managing natural resources has just begun.

3.1.1 Definition of System Profiles

Figure 3.1 illustrates the conceptual approach proposed for practical urban geological applications. The principal approach includes local investigations in the context of regional urban landscape development and consists of the following iterative procedures (Epting et al. 2008):

1. Delineation of the investigation area, including an inventory of all relevant environmental and anthropogenic/geotechnical *boundary conditions* also for adjacent zones of possible interferences or negative impacts.

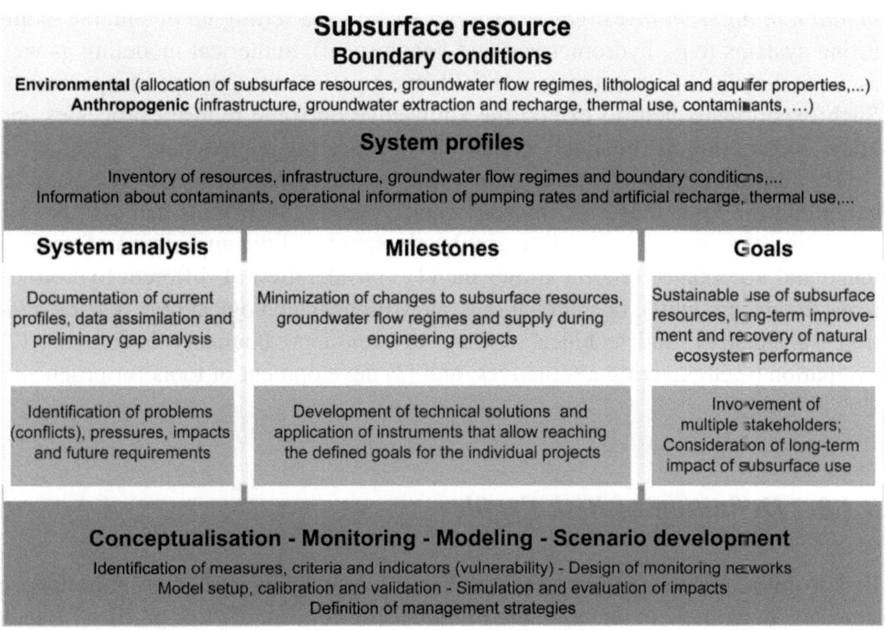

Fig. 3.1 The "system profile" concept

2. Definition of *system profiles* that describe the environmental system before, during, and after environmental changes. The concluding profile comprises the general goals for the desired future development of the system. Profiles allow decision makers to see the impact of past and present modification patterns of the system.
3. *System analysis*, including the documentation of the current system profile as well as stationary or nonstationary processes.
4. Definition of *milestones*, which represent moments when available knowledge is evaluated with respect to decisions and based on previously defined criteria. This includes the minimization of qualitative or quantitative changes of surface waters and subsurface resources, safeguarding water quality measures during water engineering projects, as well as the development of technical solutions that guarantee sustainable development of the subsurface.
5. The formulation of *goals* for a sustainable development, including sustainable use of subsurface resources, and taking into account long-term impact of geotechnical measures and future changes in usage. Whereas goals focus on a long-term sustainable development, milestones center on short-term subsurface resource protection and geotechnical issues.

The conceptual approach (*conceptualization*) incorporates the different geological, hydrogeological, and geotechnical information as well as the determination of the relevant parameters. It is accomplished by combining instruments that facilitate the adequate identification of the influences of the various single impacts on the complete system.

The core elements of this adaptive procedure include the *integration and combination of different investigative methods* such as the setting up of suitable monitoring systems (e.g., hydrometrical and geophysical), numerical modeling as well as the realization of experiments and field investigations to study specific processes. Such experiments help to extend the knowledge on some relevant processes and allow constricting the boundary conditions of modeling approaches.

The developed tools allow monitoring the relevant processes as well as extending the knowledge on environmental systems. Such tools can also have a predictive character. Finally, scenario development and the application of equivalence and acceptance criteria allows the (1) consideration of different hydrologic (e.g., floods, droughts), operational (e.g., new usages), geological (e.g., karst evolution, subsidence), and technical (effects of measures) boundary conditions; (2) evaluation of engineering alternatives; and (3) development of tools for prognosis.

3.1.2 Definition of Risk Profiles

Risk is defined as the product of the probability or frequency of an event to happen times the magnitude of damage. We introduce the concept of "risk profiles," which includes the determination of principal hazards and related damage scenarios for subsurface resources or different water supplies. The concept comprises quantitative and qualitative degradations of urban subsurface resources and the potential impact or damage for water supplies and consumers. Figure 3.2 illustrates the concept for urban environments. The figure summarizes the spatiotemporal interrelationships of natural and man-made hazards.

Natural hazards in urban areas include, e.g., earthquakes, landslides, or flood events (Sect. 5.6). However, in urban areas man-made hazards can also arise from, e.g., engineering projects (Sect. 5.2), contaminated sites (Sect. 5.3), induced subsidence processes (Sect. 5.4), or the use of geothermal energy (Sects. 5.5 and 5.6).

Together with the development of concepts for risk assessment and vulnerability (Sect. 3.3), which include the identification and characterization of risk situations, calibrated models allow to set up risk scenarios. As risk estimates are site specific, time variant, and uncertain, the determination of "acceptable" risk levels is fundamental to any management program.

The identification and localization of the relevant processes that lead to specific risk situations is the basis for differentiated subsurface resource protection measures. Selected methods used to assess and manage the risks for subsurface resources thereby are based on scientific data and provide a rational basis for quantifying hazards. In Chap. 5, we illustrate the applications of the "risk profiles" concept for several case studies. Based on preliminary risk assessment, specific risk profiles and -levels for the investigated subsurface resources and groundwater bodies are defined. Some of the specific questions that we address in the various chapters and in the process of risk evaluation include:

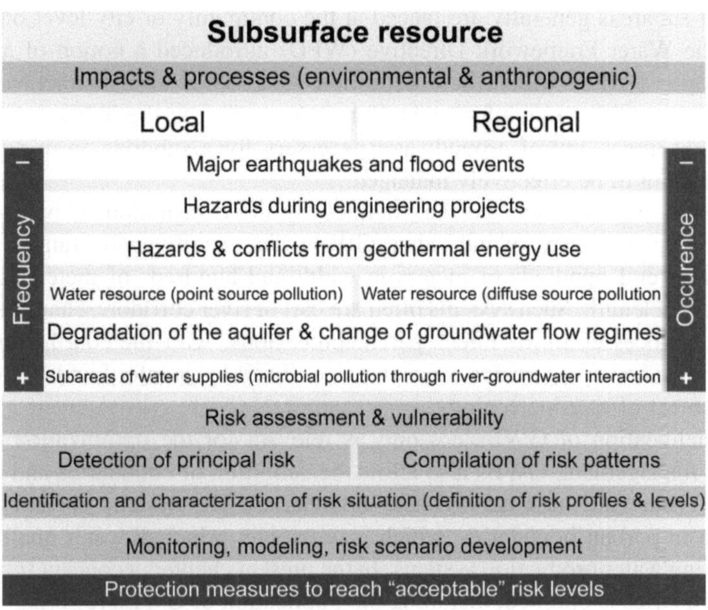

Fig. 3.2 The "risk profile" concept

1. How can risk situations for subsurface and water resources be minimized during and after infrastructure development or other human activities based on preliminary risk assessment and the implementation of adequate monitoring and modeling approaches (Chap. 5)?
2. How far can water supply systems be optimized regarding to the temporal and spatial transient character of pollution hazards and vulnerability (microbial or persistent chemical compounds; Sects. 5.1–5.3)?
3. Which are the main human and natural caused hazards to the present environmental systems and what is their frequency (including natural hazards and anthropogenic pollution; Sects. 5.3–5.6)?
4. How can a series of local water supply and thermal groundwater use systems be integrated into a supply network, based on local and regional scale risk assessment and considering long- and medium-term development of subsurface use (Sect. 5.5)?

3.2 Flow Across Boundaries

Jannis Epting and Peter Huggenberger

The management of resources in urban areas implies a definition of manageable units of the subsurface. Whereas water resource management in rural areas often is realized within large catchments or subcatchments in urban areas decision procedures for subareas generally are placed at the community or city level on a lower scale. The Water Framework Directive (WFD) introduced a notion of a *body of groundwater* (*GWB*) with a new definition – "a GWB is a distinct volume of groundwater within an aquifer or aquifers." According to the WFD, a GWB will be a management unit of groundwater necessary for a subdivision of aquifers in order for them to be effectively managed.

The size of GWBs can be, i.e., defined by the distribution of bedrock steps separating river reaches or individual GWBs. The size typically ranges between tens and several hundreds of meters in width and hundreds of meters to several kilometers in length. Such systems often are part of river corridors, characterized by exchange processes in all three spatial dimensions, with the fourth dimensions being time. The river corridor concept was outlined by Stanford and Ward (1993) in the context of riverine ecology.

The delineation of GWBs not only is relevant for the optimization of water resource management. GWBs also allow the definition of boundaries and to derive fluxes of heat and mass across these boundaries. The system in- and outflows represent important boundaries, which may explain principal water quality issues for drinking water production systems. In the present chapter, a concept for defining such boundaries is outlined, including the implication of GWBs for urban resource management.

3.2.1 River Landscape Development

Riverine landscape development is one key for understanding the formation of GWBs. The exchange between surface waters and especially rivers with groundwater are important components of GWB systems. In the last two centuries, land use and segmentation changed riverine landscapes considerably. These changes involved the straightening of rivers resulting in drastically impacts to alluvial systems and associated GWBs.

Originally, natural rivers did not follow sharply bounded lines and the boundaries between the distinct landscape elements were complex and transient. The elements within these ancient floodplains included, beside the different river terraces, sporadically inundated areas and the active channel belt, mainly consisting of coarse gravel. In the last two centuries, the former connection between the various landscape elements has been transformed into simple geometric units with well-defined boundaries. This separation into compartments strongly influenced the economic development and urbanization in the alluvial valleys in most European countries.

Accompanied by technical developments in agriculture, a series of environmental problems such as flooding, groundwater pollution, and ecological changes, including the decrease of characteristic habitats of riverine landscapes, were created within the last two centuries. However, the most fundamental changes took place in the fast developing urban areas that frequently are located in alluvial valleys. At the same time, important groundwater reservoirs are located in these areas, which are endangered by intense industrial activities and numerous traffic lines. Additionally, the high permeability of fluvial sediments (esp. fluvial shaped aquifers of the perialpine type), the frequently observed lack of protective cover layers and altered exchange processes with surface waters result in a high vulnerability of urban groundwater resources.

3.2.2 Major Interfaces

For adaptive water resource management and in order to predict changes to GWBs, the relevant influencing factors as inflows and outflows to the system have to be known. The identification of relevant processes within GWBs can occur iteratively by defining goals for the improvement of groundwater systems and formulating the relevant questions. To answer these questions, which, e.g., are related to land use or quantitative and qualitative issues of water resources, scenarios can be developed based on specific field investigations, groundwater monitoring, and numerical models. A further important element of adaptive water resource management is the evaluation of uncertainty. This approach corresponds to "work in progress" (National Research Council 2004) and was conceptualized in Sect. 3 1.

Fluxes across the different interfaces in natural coarse fluvial systems are subject to continuous dynamics not only involving water budgets, water quality, and flow patterns, but also to changing system boundaries. To understand the fluxes across interfaces in riverine landscapes our investigations focus on the evaluation of transient hydraulic boundary conditions, including the transient character of riverbed permeability as well as qualitative aspects of surface waters and groundwater (Affolter et al. 2010).

Because of the importance of boundary fluxes for the recharge of GWBs we address the major inflows and outflows to the urban groundwater system of Basel in this chapter, including (1) the role of areal groundwater recharge by infiltrating precipitation water; (2) linear groundwater recharge and discharge by river–groundwater interaction; (3) the role of boundary fluxes from adjacent catchment subareas; (4) the effect of small- to large-scale groundwater extraction for water supply, cooling, and industrial processes; as well as (5) different systems of artificial groundwater recharge.

It is documented for many floodplains, as e.g., for the Basel area, that groundwater recharge is dominated by river water infiltration, subsurface fluxes from the adjacent hill slopes and valleys, as well as artificial groundwater recharge (INTERREG III Project "MoNit" 2006). Subsequently, some general concepts about flow and thermal budgets across urban GWBs are addressed.

3.2.2.1 Areal Groundwater Recharge

The following discussion is based on data that are derived from the transnational INTERREG III Project "MoNit" (2006) "modeling of nitrate impact to the groundwater in the Upper Rhine Graben." The regional hydrological data were processed by the Institute of Hydrology in Freiburg (IHF Germany) and GIT HydrosS Consult GmbH. According to this approach, the amount of groundwater recharge was simulated that originates from percolating precipitation water. The results allow describing the spatiotemporal transient character of areal groundwater recharge within the investigation area including the degree of surface sealing. Figure 3.3 illustrates the areal distribution of groundwater recharge for the Basel area for February and August 2002. For both months it can be observed that recharge in the urbanized areas is lower than in the adjacent open space areas.

Whereas in spring, groundwater recharge is generally high, and at the end of summer, groundwater recharge is at its minimum. In summer months, areal groundwater recharge usually is lower as evapotranspiration accounts for considerable water losses to the atmosphere. Groundwater recharge is higher in the winter, illustrated by February 2002, as evapotranspiration losses are significantly lower than in summer. It can also be observed that there are major differences in recharge over short distances and that the highest recharge rates occur on the adjacent topographic elevations. This can be attributed to less urbanized areas in these regions and the altitude effect of precipitation. Also, the effect of more or less urbanized areas can be observed. Whereas areal groundwater recharge on the exhibition square (compare Fig. 3.4) is the lowest observed throughout the year, in parks

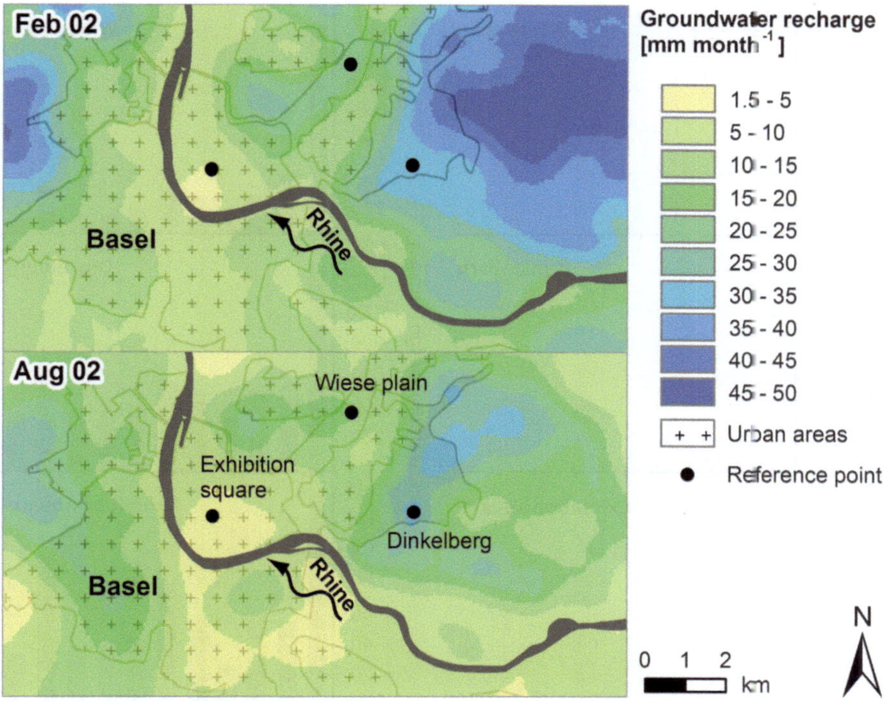

Fig. 3.3 Areal groundwater recharge in the Basel region

the amount of areal groundwater recharge can be comparable to the landscape areas.

Figure 3.4 illustrates the temporal development of recharge for three different locations. The different data sets illustrate the distribution of annual and interannual groundwater recharge for the time period 1985 to 2003 for (1) the Wiese flood plain, which is characterized mostly by open space; (2) the Dinkelberg, exemplifying the influence of altitude on groundwater recharge patterns; and (3) the exhibition square in Basel, which is characterized by a low permeable surface. Annual and interannual differences show the very variable temporal character of groundwater recharge. In highly urbanized areas, as on the exhibition square, groundwater recharge is reduced by a factor of 10 compared to open space areas, where groundwater recharge during single summer month can dominate the annual recharge.

The general distribution of areal groundwater recharge can be observed for the entire time period. However, there can be considerable differences, both seasonally and interannually. Hence, next to spatial heterogeneous recharge patterns, further temporal variations also have to be considered, especially when investigating fluxes into and out of contaminated sites.

The assessment of the spatiotemporal highly transient character of areal groundwater recharge is prerequisite to define the upper boundary conditions of high-

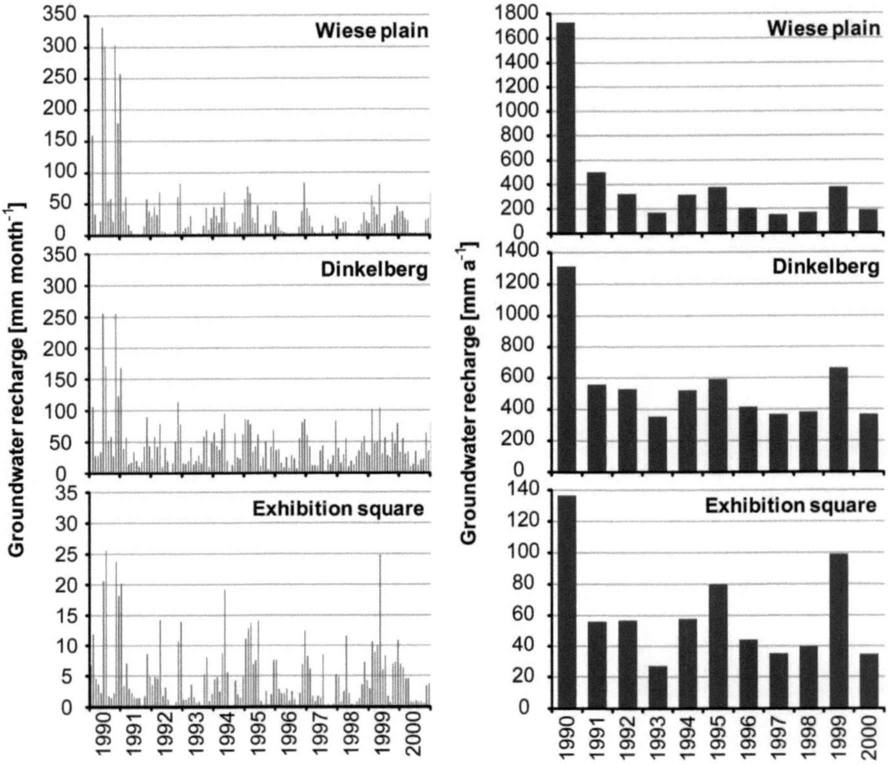

Fig. 3.4 *Left*: monthly groundwater recharge. *Right*: Annual groundwater recharge in the Basel region (see Fig. 3.3 for locations)

resolution groundwater models. Also the knowledge of recharge processes allows environmental agencies to conduct targeted measures, such as enhanced rain water infiltration, within specific areas.

3.2.2.2 River–Groundwater Interaction

Groundwater recharge in river valleys often is dominated by river water infiltration. We illustrate that the consideration of the transient character of riverbed permeability is an essential factor to be considered for quantitative hydrological and hydrogeological aspects of river–groundwater interaction. Therefore, one focus of our work is placed on the evaluation of transient hydraulic boundary conditions, including riverbed permeability and qualitative aspects of surface waters and groundwater. Whereas river–groundwater interaction investigations often are limited to site-specific issues, we developed methods that allow capturing the 4D nature of the processes (3D in space and 1D in time) over wider river stretches.

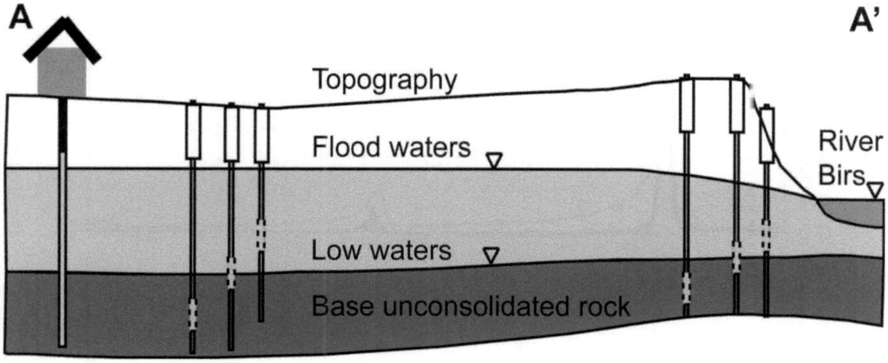

Fig. 3.5 Setup of an observation system that facilitates the measurement of groundwater parameters at different depths (see Fig. 3.8 for location of transect)

The interaction is subject to continuous dynamics involving water budgets, water quality, and flow patterns. Sediment erosion as well as transport and deposition processes is influenced by rivers that are able to exert their natural dynamics. As a consequence, in natural systems the variance of the riverbed permeability is increased temporarily, influencing infiltration rates and groundwater mixing ratios, as well as residence times for groundwater from different provenance. Within this context, effects and phenomena of river–groundwater interaction for channelized and nonchannelized surface waters are investigated in numerous projects worldwide.

To capture the transient character of river–groundwater interaction, different methods may be applied and combined, including:

1. The setup of observation systems, e.g., for depth-oriented groundwater measurements near in- or exfiltrating river segments (Fig. 3.5) allows to capture riverine filter capacities and mixing of groundwater components.
2. The realization of specific field investigations, as hydrogeophysical methods or event-oriented microbiological sampling allow (a) studying the transient

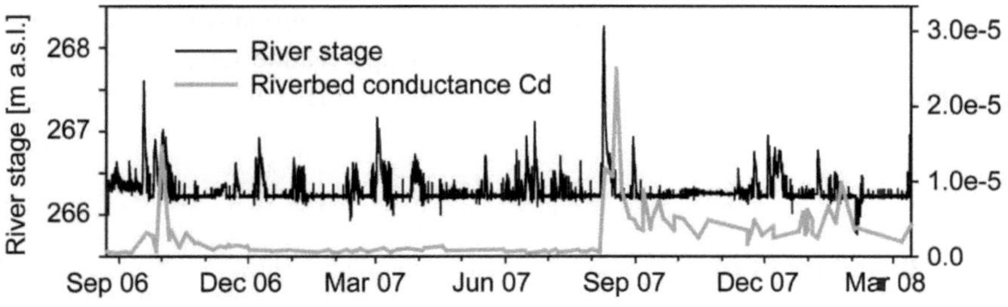

Fig. 3.6 Conductance model of the riverbed, derived from temperature data analysis. Results illustrate the temporal heterogeneous character of river–groundwater interaction and the effect of a major flood event on river bed conductance

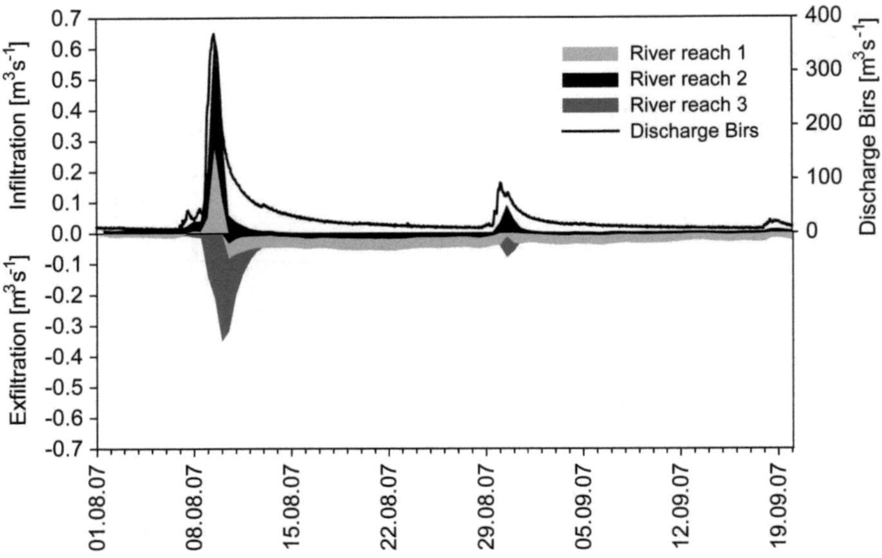

Fig. 3.7 Modeling results illustrating the spatial heterogeneous distribution of river water infiltration and groundwater exfiltration for different river reaches (see Fig. 3.8 for location of reaches)

character of relevant processes and (b) extrapolating 1D information of measurements into 2D zones of enhanced or reduced river–groundwater interaction (Sects. 4.2 and 4.3).

3. The simulation of the transient character of river bed conductance by empirical or physically based methods. Such methods allow quantifying transient fluxes across river beds and river banks. Subsequent, conductance models can be integrated into groundwater flow and transport models. Figure 3.6 illustrates a conductance model for the riverbed that was derived from temperature data analysis (see Sects. 5.2 and 5.4 for project details).
4. 3D groundwater flow and transport modeling, in order to (a) derive areal groundwater flow budgets and residence times and to (b) facilitate the definition of scenarios with changing hydrological and technical constraints (Sect. 5.1). Scenario techniques then allow studying river–groundwater interaction processes at regional and local scales (e.g., Upper Rhine Graben, individual catchment areas, river reaches, and capture zones of extraction wells; Fig. 3.7).

3.2.2.3 Boundary Fluxes

The knowledge of the relevant lateral boundary fluxes is of key importance for an effective management of GWBs and long-term groundwater protection. The main lateral boundary fluxes are the inflows and outflows of the main valleys

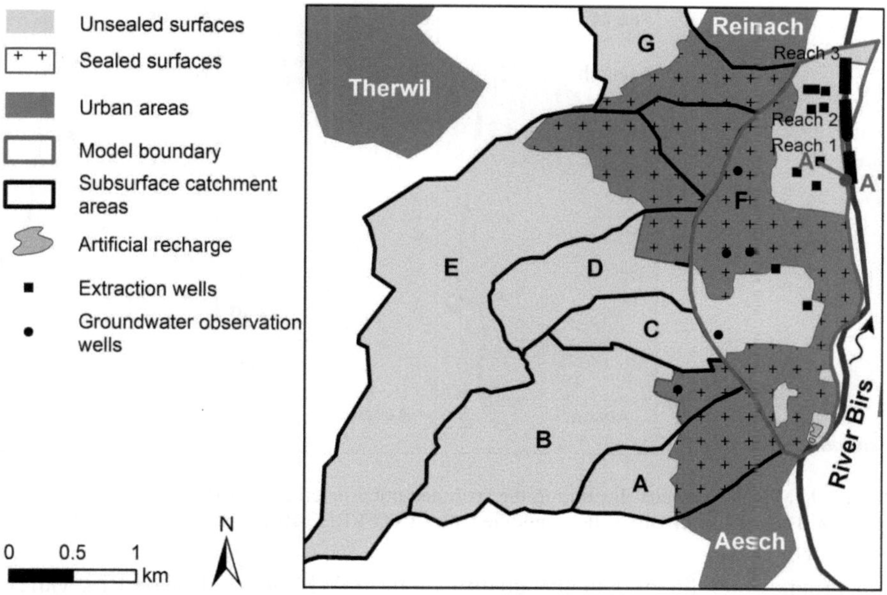

Fig. 3.8 Subsurface catchment areas to the GWB of the Lower Birs valley and boundary segments (modified after INTERREG III Project "MoNit" 2006)

and those from the hill slope catchments. Storage capacity, the groundwater flow regime (residence times, budgets across different cross-sections), and water quality strongly depend on these lateral fluxes and the water quality entering the GWBs at the different boundaries.

For modeling approaches, main valley inflows and outflows often are considered as specified head boundaries, as groundwater observation wells often are concentrated in the main valleys. Inflow from adjacent catchment areas generally is more diffuse and therefore often is considered as specified flux boundary.

In Fig. 3.8, we illustrate how the inflow to the GWB of the Lower Birs valley from the adjacent catchment areas was simulated. In a first step, we evaluated the surface and subsurface catchment areas that drain into the boundary of the GWB by surface modeling of digital elevation models of the surface and the bedrock (cf. Fig. 4.3). Several subsurface catchment areas are illustrated. In a next step, the various areas were combined with transient groundwater recharge data derived from the method described earlier. Furthermore, for areal groundwater recharge we delineated more or less urbanized areas.

The analyses show that the subsurface catchments drain via segments into the GWB. Whereas some catchments are considerably small but constitute to a large boundary segment of the GWB, others are large and drain via small segments to the GWB. This concept allowed us to estimate fluxes of different compounds, such as nitrate, from the agricultural areas into the GWB. Furthermore, this approach allowed us to evaluate the influence of measures in the different catchment areas

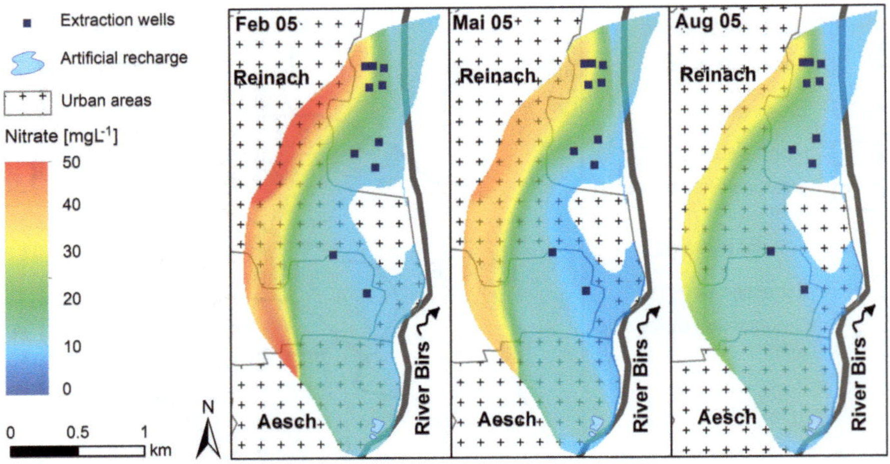

Fig. 3.9 Modeling results illustrating the spatiotemporal heterogonous distribution of nitrate in the GWB of the Lower Birs valley (modified after INTERREG III Project "MoNit" 2006)

on groundwater fluxes and the quality of the GWB of the Lower Birs valley. Figure 3.9 shows the results of a nitrate transport model and the annual development of nitrate inflow to the GWB. Most inflow occurs in spring via catchments B and E which are considerably large. However, the inflow occurs via small boundary segments to the water body. The effect of infiltrating river water with comparable low nitrate concentrations is observable and more pronounced during flood events. The groundwater extraction wells closest to the western boundary of the groundwater body are influenced by high nitrate concentrations in spring. This kind of approaches allowed us to identify the source of contamination and thus to define effective measures within the catchment areas that substitute the highest nitrate influx to the groundwater body; in this case in catchments B and E.

3.2.2.4 Groundwater Use and Artificial Groundwater Recharge

Groundwater in the urban area of Basel is intensively used. Besides several medium- to large-scale municipal water supplies, industry uses groundwater for cooling and production. To guarantee the demands, surface water is recharged to the groundwater. In particular in the Basel area, artificial groundwater recharge contributes substantially to the overall groundwater budget. Different artificial recharge methods are applied: (1) injection wells and infiltration ditches, as utilized in the Hardwald; (2) artificial infiltration ponds with filtering layers, as utilized in the lower Birs valley; and (3) infiltration fields, as utilized in the Wiese flood plain (Fig. 3.10).

There are three main reasons for the need of artificial recharge: (1) groundwater uptake exceeds the natural groundwater recharge; (2) by using previously filtered, and to some degree treated river water, the quality of extracted water can be improved; and (3) the setup of artificial hydraulic barriers allows restricting the

Fig. 3.10 Different methods for artificial groundwater recharge in the Basel region

inflow from boundaries with poor water quality. In the Basel area, a combination of these three factors is the main reason for maintaining the present management scheme.

3.2.2.5 Flow Budgets of Urban GWBs

Figure 3.11 illustrates the distribution of flow budgets to different GWBs in the Basel area. River water infiltration, inflow from the adjacent catchments, and artificial groundwater recharge are the main contributions to groundwater renewal in the investigated GWBs.

Groundwater recharge for the three different GWBs by infiltrating river water can reach up to 12 to 23% of the total flow budget (in and out). Groundwater recharge through infiltrating river water especially for the Wiese plain and the Birs valley therefore are substantial components of overall recharge. Obviously, groundwater recharge is not dominated by percolating precipitation water. Also the influx from the adjacent catchments is, except for the Birs valley, a comparably small component. The data clearly document the importance of good quality surface waters for a long-term protection of the groundwater quality.

3.2.2.6 Thermal Budgets of Urban GWBs

The growing interest to use water resources for heating or in most urban areas for cooling requires an analysis of the thermal state of GWBs.

The temperature of noninfluenced groundwater in near surface aquifers corresponds time-delayed and damped to the progression of air temperature. Longer residence times and increasing depths within the aquifer result in greater time delays and damping. Temperature of groundwater with long residence times of more than 1 year corresponds to the longtime annual average of air temperature.

In the Basel area, the main aquifers consist of highly permeable gravels. Some of the diffuse boundary fluxes from the adjacent hill slopes to these aquifers are characterized by long residence times. However, in the vicinity of rivers groundwater recharge and thermal regimes are dominated by short-term changes and the processes of river–groundwater interaction (see above). In order to understand the

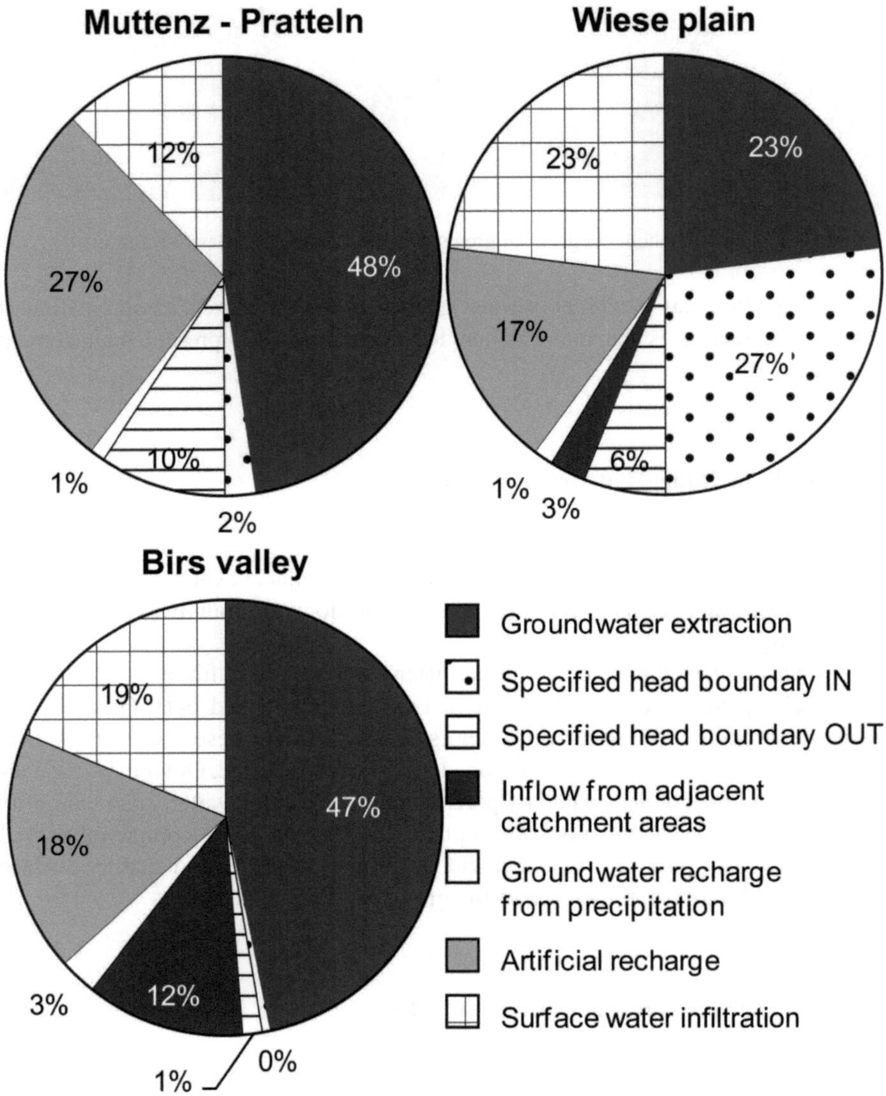

Fig. 3.11 Flow budgets of several GWBs in the Basel region. The locations of the investigated GWBs are illustrated in Fig. 4.10

consequence of anthropogenic influences as well as climate change (see Sect. 3.4), different areas within an urban GWB have to be evaluated separately. In Sects. 3.4 and 4.5, an example is illustrated how different areas of an urban GWB can be delineated based on multiparameter statistics.

In case anthropogenic impacts on urban GWB are not reduced or even impaired, a long-term increase of groundwater temperatures will occur. However, in urban areas, climate change effects will be secondary (see Sect. 3.4). More pronounced are anthropogenic impacts and changes of the use of aquifer resources. In case no measures will be taken against this development, groundwater temperatures in the Basel area will rise aerially up to 14 to 18°C. These changes can differ regionally in the vicinity of rivers, artificial recharge, and groundwater use.

The spatiotemporal localization and the magnitude of changes generally cannot easily be evaluated as there are not enough data and considerable uncertainties. To adequately detect and evaluate the influences of the different changes on urban groundwater resources, it is necessary to condition existing long-term data sets. Further it is necessary to evaluate the influence of long-term annual, seasonal, and event driven developments of groundwater temperatures for different zones of urban GWBs. This also includes an assessment of the progression of groundwater temperatures for different depth of the aquifer within these zones. To capture specific issues of climate and anthropogenic change, supplementary observation stations should include specific and precise monitoring devices for head, temperature, and also oxygen measurements. Therefore, the goal for the next couple of years is the implementation of a series of robust instruments that allow evaluating and differentiating the various boundary fluxes to urban GWB.

In Sect. 5.5, we introduce tools for adaptive thermal groundwater management in urban areas. The results from this management approach and the evaluated scenarios allow estimating groundwater fluxes and residence times for different urban areas. Such knowledge is the base to evaluate the impact of existing and new thermal groundwater users and to define new policies for the thermal use of the subsurface. This also includes management concepts for individual groundwater users and the provision of guidelines for subsurface constructions that reach into the groundwater. Measures to minimize the anthropogenic influences and the heating of groundwater could include the isolation of cellars and the minimization and quantification of thermal recharge.

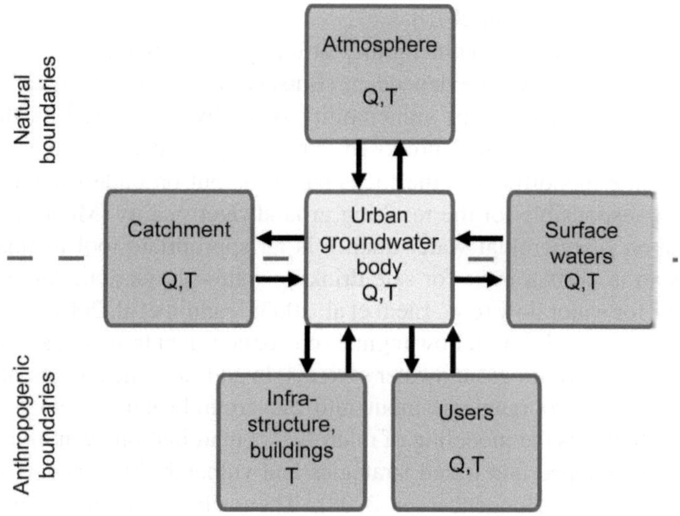

Fig. 3.12 Fluxes in and out of an urban groundwater body

In the vicinity of national and international boarders, the management of groundwater resources today and in the future is subject to agreements of the main groundwater users. Such agreements require a discussion on long-term goals for the sustainable development of groundwater resources. Next to more restrictive approval practices, legal aspects should base on the demand that thermal and water budgets are balanced (Fig. 3.12). This suggests combined thermal use for heating and cooling and includes financial incentives.

3.3 Vulnerability and Quality Control Systems

Christoph Butscher, Peter Huggenberger, Stefan Scheidler,
Annette Affolter, and Jannis Epting

The subsurface of many urban areas include karst terrains or highly conductive fluvial deposits. Such areas are most productive and provide a relevant amount of drinking water all over the world. However, these groundwater resources are known for being very susceptible to pathogen contamination especially after precipitation events due to point recharge and river–groundwater interactions. Rapid water quality change often can be attributed to aquifer heterogeneity. This includes the heterogeneous structure of karst and highly permeable gravel aquifers which both cause slow and fast water responses observed in springs and in hydrographs near rivers.

Accounting for these special permeability characteristics, different methods have been developed to understand the vulnerability of these systems. In this chapter, we link the outcome of some novel methods with the methods of the management of drinking water quality as outlined by the WHO drinking water guidelines (Water for health 2010).

In natural flow systems vulnerability strongly depends on the recharge conditions, which are highly time dependent (Butscher and Huggenberger 2009). The resulting temporal variation of vulnerability is hardly considered by the present vulnerability mapping methods proposed to outline groundwater protection zones.

To overcome this difficulty, the focus has to be put on understanding the processes being responsible for the resulting groundwater quality. Modeling the temporal variation of microbial water quality is an appropriate tool to this end and constitutes an important issue for safe drinking water production. Although many models exist for water flow (e.g., Liedl et al. 2003; Jeannin et al. 2001) the effects of interactions between different flow regimes on contaminant transport are only partly known in karst and river–groundwater systems. In addition, the amount and temporal variation of microorganisms input into the groundwater system is often not known, which makes the modeling of microbial contamination even more difficult.

We have developed and tested strategies and vulnerability concepts in the last years (Butscher and Huggenberger 2008). They allow examining groundwater resources from a quantitative and dynamic point of view and facilitate a clear identification and establishment of reference criteria for quantification, comparison, and validation purposes (Brouyère 2003). In the first part of this chapter, we present methods for vulnerability assessment including a regional planning concept for the sustainable use and protection of water resources in urban areas. The approach allows both spatial and temporal aspects of groundwater vulnerability to be addressed.

Reliance on water quality determination alone is not sufficient to protect public health. As it is neither physically nor economically feasible to test for all drinking water quality parameters, the use of monitoring effort and resources should be carefully planned and directed at significant or key characteristics (WHO, Guidelines for drinking-water quality). The control of the microbial and chemical quality of drinking water requires the development of management plans, which provide the basis for system protection and process control to ensure that numbers of pathogens and concentrations of chemicals present a negligible risk to public health and that water is acceptable to consumers.

The present approach of the WHO drinking water guidelines to drinking-water supply management includes a systematic assessment of risks throughout a drinking-water supply – from the catchment and its source water through to the consumer – and identifies the ways in which these risks can be managed. This also includes methods to ensure that control measures are working effectively. The setup of management systems based on the Hazard Analysis and Critical Control Points (CCPs) concept (HACCP; e.g., Rauch 2009) incorporates strategies to deal with day-to-day management of water quality, including upsets and failures. Therefore, in the second part of this chapter we discuss the role of quality control systems, which include the monitoring of physical, chemical, and microbiological parameters, the definition of CCPs as well as flux calculations, which can be derived from groundwater modeling.

3.3.1 Vulnerability Assessment Methods

The term vulnerability we use in this chapter applies to the definition of Vrba and Zoporozec (1994), who defined vulnerability as an intrinsic property of a groundwater system that depends on the sensitivity of that system to human and/or natural impact. There are basically two concepts of groundwater vulnerability: intrinsic and specific vulnerability. Intrinsic vulnerability refers to the natural, geogenic vulnerability without relation to a specific contaminant or contaminant source. In contrast, specific vulnerability considers human activities and the impact of a particular land use. It is related to a specific contaminant or contaminant source (Vrba and Zoporozec 1994). Since the introduction of the concept of groundwater vulnerability, many strategies have been developed for vulnerability assessment and groundwater resource management. The proposed methods have also been integrated into environmental policy and regulation practices (National Research Council 1993; European Water Directive 2000, etc.), most of them placing an emphasis on intrinsic vulnerability.

Vulnerability assessment methods can be divided into "source" and "resource protection" methods (Hötzl 1996). Resource protection methods aim to protect all of the groundwater, whereas source protection methods focus on the protection of a discrete water source. Source protection methods are based on the assumption that some places within a catchment area are more vulnerable to contamination than others.

Since the mid-1980s, worldwide researchers have developed various strategies for the protection of groundwater sources, with methods for groundwater vulnerability mapping as one of the most important (National Research Council 1993). Mapping methods are based on the assumption that some places within the catchment area of the regarded groundwater source are more vulnerable to contamination than others. Most mapping methods are multiparameter methods where different criteria (e.g., protective cover, flow velocities, etc.) are used to estimate the vulnerability of the groundwater in a certain area. The criteria are attributed an index value, and the index values of the different criteria are used to calculate a protection factor at every place in the catchment. An overview of vulnerability mapping methods is given by Vrba and Zoporozec (1994).

However, there are problems with the protection of groundwater sources that cannot be solved by mapping methods. First, mapping approaches hardly consider the time dependences of vulnerability, though many studies show that groundwater quality is a function of time. Second, the various indices used to generate vulnerability maps are largely empirical. There is a need to examine vulnerability from a quantitative point of view. Third, though vulnerability maps take important characteristics of groundwater catchments into account, they are mostly not adequate to explain the actual processes taking place. And last, the existing mapping techniques indicate the evaluated groundwater vulnerability solely in the catchment areas. However, an indication of the vulnerability of a particular groundwater source would be a major aid to drinking water suppliers and regional planners.

In the last years, we have developed and tested an integrative concept that allows examining groundwater vulnerability from a quantitative and dynamic point of view (Butscher and Huggenberger 2009). The concept is based on a combination of mapping with groundwater modeling methods, the latter facilitating the identification and establishment of reference criteria for quantification, comparison, and validation purposes. It was developed for karst areas, but can also be applied to GWBs in urban areas. The approach includes the following steps (Fig. 3.13): (1) delineation of GWBs; (2) field measurements, e.g., pumping and tracer tests to provide information on flow velocities and aquifer properties; (3) development of calibrated groundwater models and calculation of vulnerability indices based on model results; (4) traditional vulnerability mapping; and (5) vulnerability maps established based on determined GWB, results of field investigations, groundwater modeling, and vulnerability mapping (integration of methods). In the last step, the static information on the vulnerability distribution in the catchment, derived by vulnerability mapping, is complemented by quantitative information on the vulnerability of a particular groundwater source and its temporal variation derived from field experiments and numerical modeling.

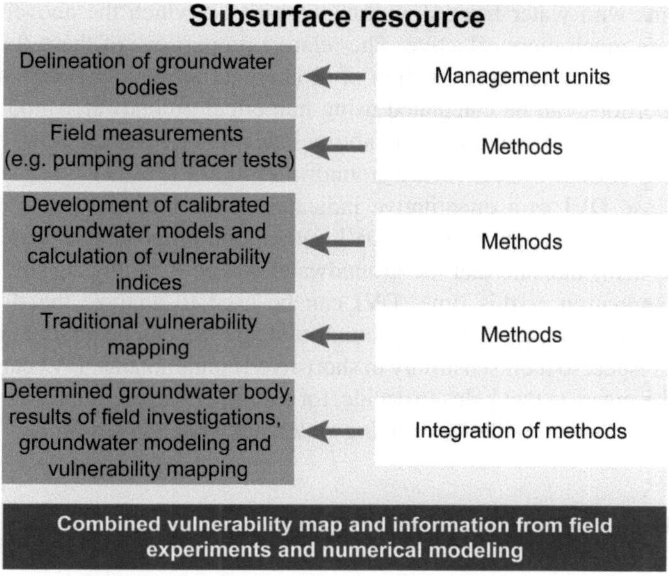

Fig. 3.13 Conceptual approach proposed for integrative vulnerability assessment. The approach concludes with a combined vulnerability map and time series providing quantitative spatial and temporal information (modified after Butscher and Huggenberger 2009)

3.3.1.1 Use of Groundwater Models for Vulnerability Assessment

In this section, we want to outline how calibrated numerical groundwater models can be used for vulnerability assessment. Before doing so, we first have to introduce a "dual" concept of vulnerability. Many groundwater systems include fast and slow flow systems. For example, in karst springs, conduit flow and rock matrix flow components can be distinguished, and in groundwater wells in porous aquifers close to infiltrating rivers, base flow and event flow (flood) components can be recognized. In addition, contaminants can be classified as short-lived or persistent. The vulnerability of a groundwater source to short-lived and persistent contaminants is expected to be different and will depend on variable proportions of slow and fast flow systems in the groundwater source and its variation in time. Vulnerability to short-lived contamination is mainly related to fast-flow components, whereas vulnerability to persistent contaminants is also influenced by slow-flow components.

3.3.1.2 Vulnerability to Short-Lived Contamination

Fast-flow systems are especially sensitive to short-lived contamination because short travel times in fast-flow systems reduce the effectiveness of natural attenuation processes such as adsorption, degradation, and filtration. The sensitivity of a groundwater source to short-lived contamination is reduced by dilution of water in the fast-flow systems with water from slow-flow systems, in which the above-mentioned processes are much more effective. The relative proportions of these flow systems in a groundwater source as a function of time are an intrinsic property of a GWB. These proportions can be calculated using numerical groundwater models. Therefore, we introduce the dynamic vulnerability index (DVI), defined as the ratio of the contributions of these systems in the groundwater source (i.e., DVI = fast flow/slow flow). We use DVI as a quantitative indicator of the intrinsic vulnerability of a groundwater source. High values of DVI (i.e., a high proportion of water from the fast-flow system) indicate that the groundwater source is highly sensitive to short-lived contamination at this time. DVI can be used to analyze the dynamics of vulnerability. In addition, different groundwater sources can be quantitatively compared with respect to their sensitivity to short-lived contaminants. DVI can serve as a vulnerability measure that helps to decide, for instance, which groundwater source of a certain GWB is preferable for drinking water supply at a given time.

3.3.1.3 Vulnerability to Persistent Contamination

The DVI introduced earlier can serve to characterize the dynamics of vulnerability with respect to short-lived contaminants. However, it is not a direct representation of the potential effect of a contaminant. Specifically, the vulnerability to persistent contaminants that can be stored and enriched in slow-flow systems cannot be assessed using DVI. To have an objective measure of the endangerment of groundwater quality also for persistent contamination, we introduced the dynamic vulnerability concentration (DVC). DVC is the concentration of a "standard contaminant" (SC) in a groundwater source resulting from a standardized input of SC into the GWB. In the following, we suggest two ways to use numerical groundwater models that simulate breakthrough curves (bt curves) of DVC for vulnerability assessment.

First, the input of SC is implemented continuously in the model. This is also a realistic scenario (as a simplification) for frequent contaminant inputs. A nondegradable contaminant (degradation rate equals to 0), representing an ideal tracer, is used, taking into account only the hydrological characteristics of the system independently of the quality of the contaminant. Because a degradation rate of zero already is a quality of a contaminant, additional simulations using a range of degradation rates of SC would allow an assessment of vulnerability that is influenced less by the properties of the contaminant.

Second, the input of SC is implemented as a short pulse ("Dirac input") in the model. In COST 620 (2004), Brouyére advocates the assessment of vulnerability from a quantitative point of view. He argues that the main questions to be addressed are (1) when, (2) at what level, and (3) for how long will a contamination occur at a spring or well. We calculated the first arrival time of SC, the maximum concentration of SC, and the duration of contamination in the groundwater source using numerical groundwater models. According to Brouyère (2004), the first arrival time, the maximum concentration level, and the duration of contamination represent an objective quantification of vulnerability that can be used for the validation of vulnerability maps. The calculated values can also be used to quantitatively compare the vulnerability of different groundwater sources to persistent contamination.

3.3.1.4 Benefits of Vulnerability Modeling

The proposed vulnerability assessment using numerical models could be of great interest for drinking water management. First, when spring water quality is insufficient in spite of a protection zone delineated by vulnerability mapping, the benefit of the invested mapping efforts is low. The present modeling approach, in contrast, does contribute to a better understanding of the actual groundwater system. Second, the inclusion of the dimension of time into this approach adds a degree of freedom to water protection strategies. Temporally limited land use restrictions in the catchment areas of drinking water sources and water withdrawal management are possible future applications of temporal approaches. And third, the present approach allows the comparison of different groundwater sources with respect to their sensitivity to contamination from a quantitative point of view, providing an objective basis for a regional planning of drinking water supply. However, the approach gives no information on the spatial vulnerability distribution in the GWB. Therefore, the modeling approach is a valuable complement to vulnerability mapping techniques, but the protection of a drinking water source requires a combination of both mapping and modeling methods.

3.3.1.5 Sample Application

The applicability of the concept for vulnerability assessment outlined in the above section is illustrated for an urban GWB in an alluvial (porous) aquifer that is connected to a karst aquifer. The GWB is located near to the urban agglomeration of Laufen within a Jura fold-belt, Switzerland (Fig. 3.14). The alluvial valley aquifer is used for drinking water extraction. Due to sporadic bacteriological contamination in the untreated water of an extraction well, an observation network and a high-resolution numerical groundwater model were set up. Depending on the hydrologic boundary conditions, the extracted groundwater is more or less influenced by infiltrating river water or water with short residence times within the conduit system of the karstified carbonate formations. Prior to the investigations it was assumed that infiltrating river water with short residence times is the major source of sporadic bacteriological contamination in the untreated water of the extraction well. However, hydrochemical and microbiological signatures, as well as groundwater heads revealed that river–groundwater interaction is not the only process to be considered, but that the shallow valley aquifer interacts with the regional scale karst system as described earlier (Fig. 3.15).

The first step of the approach we developed (cf. Fig. 3.13) aims at the delineation of the GWB. The delineation requires a good knowledge of the 3D geology of the

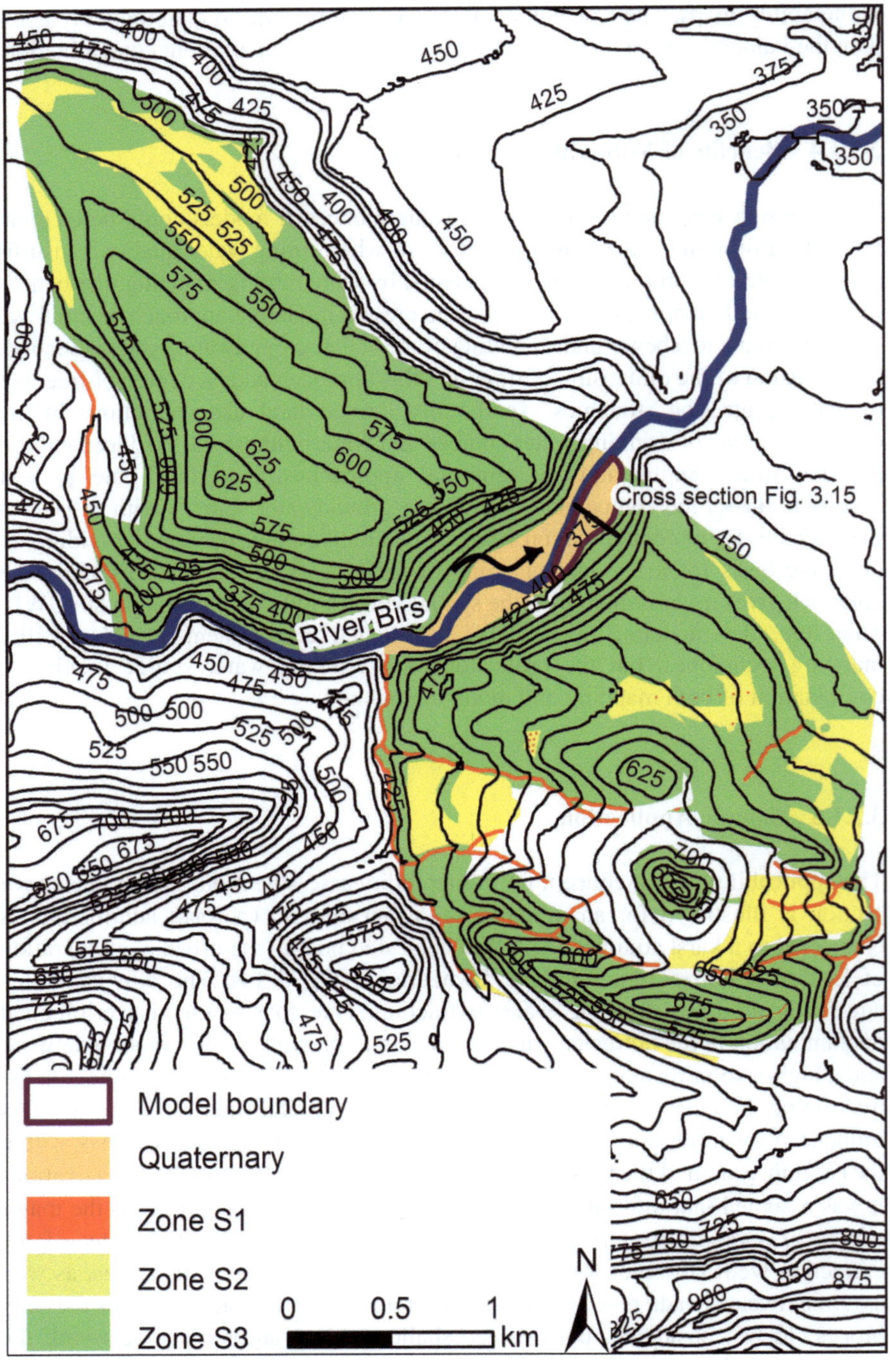

Fig. 3.14 Subsurface catchment area to the GWB used for drinking water production near the urban agglomeration of Laufen and results of the EPIK mapping method

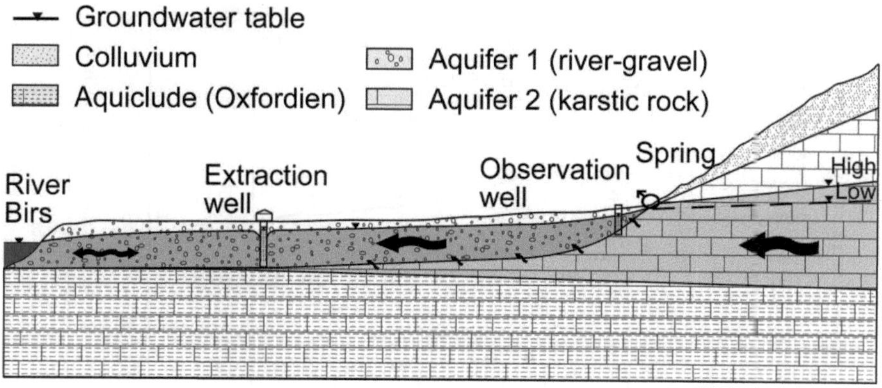

Fig. 3.15 Schematic cross-section through the GWB and the adjacent hill slope illustrating the interaction of the shallow valley aquifer with the regional scale karst system (see Fig. 3.14 for the location of the cross-section)

study area. A strategy to delineate GWBs in karst areas was proposed by Butscher and Huggenberger (2007) using the aquifer base gradient method. The method corresponds to the modeling of the aquitard surface as described in Sect. 3.2. The approach allowed us to define the subsurface catchment of the investigated drinking water well considering tectonic structures such as folds and different types of fault systems. In mature karst systems as most of the systems in the Birs valley, most of the precipitation reaches the alluvial valley aquifer via preferential flow in the conduits of the karst system. The conduit flow paths are strongly influenced by the geometry of the Oxfordian marls and clay sequences (aquitard). In non mature karst systems, inflow to alluvial valley aquifers not only is restricted to flow on the aquitard and can occur already in upper parts of the aquifer. Figure 3.14 shows the delineated catchment area of the alluvial valley aquifer including the surrounding karst system using the described approach, without considering the catchment area of the river. This catchment area is considered to represent the maximal area of influence of the drinking water well.

Within the catchment area, karst features relevant for infiltration and transport were mapped with the EPIK method (Doerfliger et al. 1999) (step 4 of the approach, cf. Fig. 3.13). The result of the EPIK mapping shows that there are no regions with a very high vulnerability that would require the most restrictive protection measures, corresponding to the groundwater protection zone S1. However, areas where karstified geologic formations outcrop have a high vulnerability and belong to the protection zone S2, where still a series of protection measures are required. Most of the catchment area is protected by low permeable cover layers and was found to have a moderate vulnerability. For the protection of groundwater in these areas, a protection zone S3 is sufficient.

A numerical groundwater model was developed and calibrated using data from field experiments (continuous measurement of river stage and hydraulic head in observation wells, tracer tests). The setup and calibration of a numerical

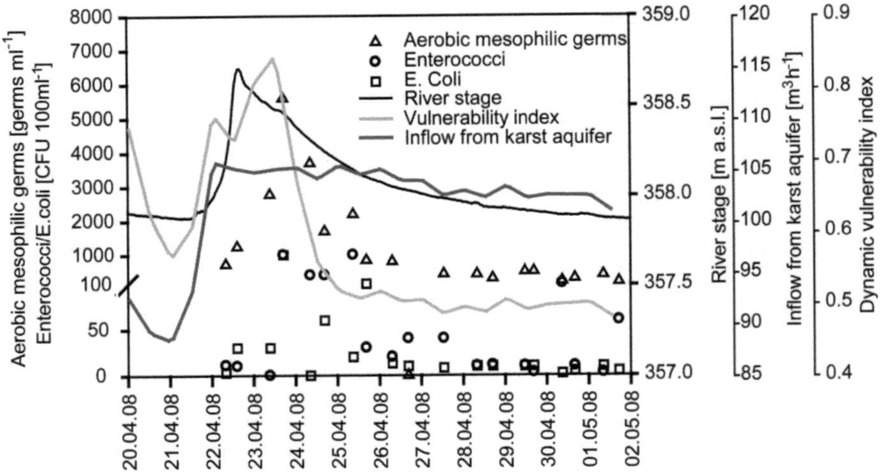

Fig. 3.16 Data set of a flood event in April 2008 illustrating the applicability of the vulnerability index in context to river discharge, measured microbiological parameters, and the modeled karst system inflow

groundwater model based on field experiments corresponds to steps 2 and 3 of the presented approach (cf. Fig. 3.13). The model was used to calculate variable contributions of different flow system to the drinking water well (slow-flow component corresponding to the base flow in the alluvial aquifer and fast-flow component corresponding to event flow originating from the karst system). Based on the calculated flow contributions and using a base flow separation method (e.g., Dyck and Peschke 1995), the DVI and its variation in time was determined for the drinking water well (step 3 in Fig. 3.13). Figure 3.16 illustrates the calculated DVI and its variation in time during a flood event in April 2008 at the river Birs in the study area. During this flood event, microbiological impact to the drinking water well during and after the flood event could be observed. The progression of DVI is in good agreement with the measured microbiological parameters. The fact that the peak in DVI precedes the discharge peak in the river and the first appearance of microbiological contamination suggests that calculated DVI can be used as an early warning parameter for drinking water supplies.

The combination of catchment area delineation, field experiments (high-resolution hydrometrical, hydrochemical, and microbiological data acquisition), groundwater modeling, and vulnerability mapping allowed the:

1. Delineation of the catchment area of the karst system contributing to the drinking water well and assessment of the vulnerability distribution in the catchment area.
2. Identification of the relevant processes that influence the vulnerability of the drinking water well using groundwater models. Results include the description of the (a) transient character of river–groundwater interaction and (b) the interaction of the shallow valley aquifer with the regional karst system for different hydrologic boundary conditions.

The results can be used for the prediction of drinking water quality and provide a decision basis for risk assessment and Adaptive Water Management (AWM) for the investigated GWB.

3.3.2 Quality Control Systems

In urban areas where the impact of human activity is high, groundwater resources increasingly contain micropollutants at low concentrations (Schirmer et al. 2007). Some contaminants can be related to human activity, such as pharmaceuticals, personal care products, and endocrine-active substances. As their investigation only started recently they are also called "emerging chemicals." The fact that some chemical compounds occur only some years after they were introduced on the market undermines the importance of comprehensive monitoring campaigns. The effects of these contaminants in the environment are widely unknown. But also other contaminants, such as microbiological pollution in groundwater extraction wells near rivers, are subject to changes, and a reliable monitoring of such contamination is essential to protect drinking water supplies.

Critical Control Points (CCP; e.g., Rauch 2009) mark moments of relevant levels concerning the quality of drinking water, where it is possible and necessary to avoid a health risk and to eliminate the risk or minimize the impacts to an acceptable level. CCPs allow to identify the risk of pollution and to adapt the control and operation system (e.g., depending on river–groundwater interaction). CCPs should be continuously measureable and each CCP should address a specific risk. Next to the easy to measure parameters as hydraulic heads, temperature, and electrical conductivity, CCPs could involve turbidity, precipitation, river stage, hydraulic gradients, and flow velocities in certain GWB regions or particles.

Based on preliminary risk assessment, specific risk profiles for GWBs can be defined (see Sect. 3.1). The definition of CCPs for selected parameters, which are continuously measured in observation and groundwater extraction wells, allows to make decisions based on observed changes and to appropriately update operation systems. Methods used to assess and manage the risks for GWBs are thus based on control data which provide a rational basis for quantifying hazards.

Groundwater protection strategies in urban areas are mainly oriented to requirements of particular sectors, e.g., protection zones for pumping wells. Currently, quality control systems for GWB often are restricted to measuring a limited set of parameters and defining thresholds on a routine basis (i.e., measurements every 3 months according to a predefined plan). Rarely, time series of data are analyzed and interpreted in the context of the relevant system processes and changing hydrological and operational boundary conditions.

Whereas during average hydrologic boundary conditions no pathogens can be found in the extracted groundwater, considerable pollution often can be observed during events. Such events can be captured through systematic event-oriented sampling. A paradigm change in sampling strategies from sampling according to predefined intervals to continuous sampling is necessary (Affolter et al. 2009). To this end,

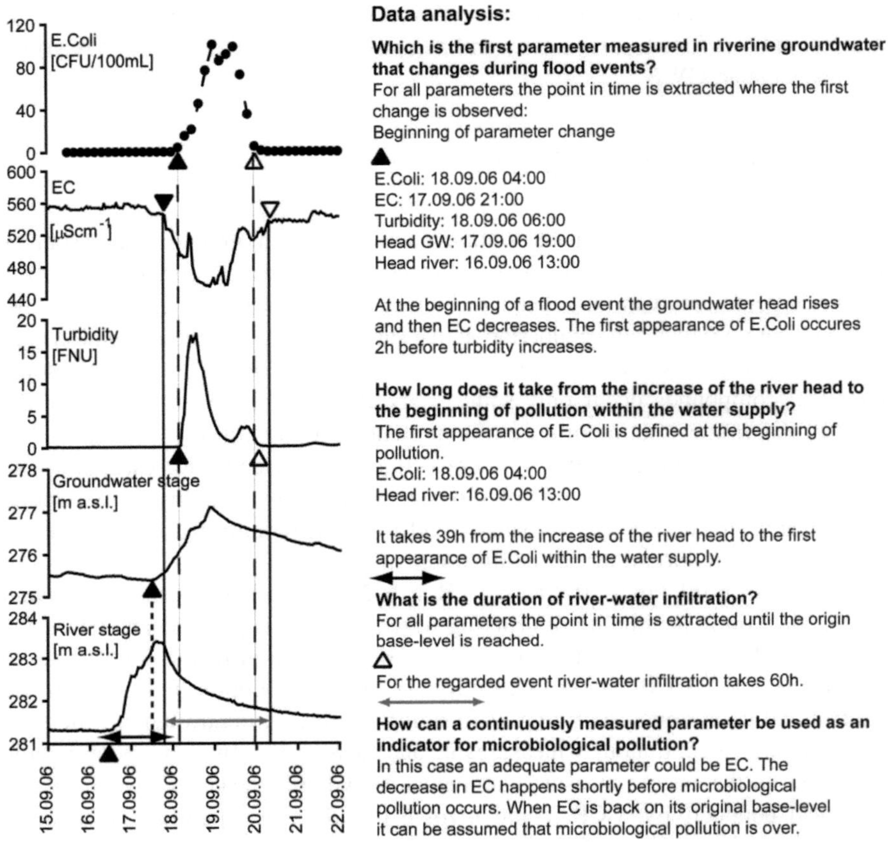

Fig. 3.17 Interrelation of system parameters. Time series of river and groundwater head as well as turbidity, EC and *E. coli* during a flood event (Affolter et al. 2009)

easy to measure parameters must be derived as CCPs for potential risk situations. Answers to the following two questions support in finding suitable CCPs:

1. Which hydraulic and/or qualitative parameter indicates potential risk situations and allows characterizing and predicting vulnerability of the urban GWB?
2. Which easy-to-measure representative physical properties can be used as proxies for more difficult to measure quantities, particularly related to water quality, and how can these proxies be used to improve interpolation between point measurements (e.g., concentrations in individual samples)?

Pronk et al. (2007) showed how particle size distribution can be used to define CCPs for karst spring water resources. Also Fig. 3.17 exemplifies the approach. In this example, electrical conductivity is an adequate parameter to announce microbiological pollution and to estimate the time period of endangerment.

The example shows that the monitoring of physical, chemical, and microbiological parameters, as well as flux calculations, which can be derived from groundwater modeling, allows assessing the vulnerability of GWBs. In this way, the impact of human activities and microbiological pollution on urban water systems and on processes within urban GWBs can be determined.

3.4 Climate Change

Jannis Epting, Peter Huggenberger, Christoph Butscher, and Jelena Simovic Rota

In the context of the ongoing debates on the impact of climate change to society, the cantonal authorities of Basel-Stadt asked us to evaluate and summarize the present state of the groundwater temperatures in the city. The evaluation included an estimation of the influence of anthropogenic and climate change to quantitative and qualitative aspects of groundwater resources. It is important to see how the vulnerability of urban GWBs is influenced by changing hydrological conditions. Although climate change is expected to have a strong impact on the hydrological cycle, little is known about its effect on groundwater in general, and on groundwater quality in particular (Bates et al. 2008). In urban areas, the influence of climate change is overlapped by the changes implied by numerous urban activities that influence groundwater temperatures, quality, and quantity.

The abrupt temperature shift in the late 1980s, which is commonly observed in various other environmental and aquatic systems, including Swiss rivers (Hari et al. 2006), is now acknowledged to be a manifestation of a large-scale climate regime shift associated with changes in the behavior of the North Atlantic Oscillation (e.g., Conversi et al. 2010).

In the first part of this chapter, we discuss several positive and negative feedback mechanisms concerning water and thermal budgets and the impacts of climate change in urban environments. In the second part, we illustrate the effects of predicted climate change on groundwater vulnerability in urban environments. In the third part, we illustrate how historical groundwater temperature data can be analyzed to delineate different zones of urban GWBs and to optimize future observation networks to capture the impacts of anthropogenic and climate change.

3.4.1 Climate Change and Feedback Mechanism in Urban Environments

Global warming due to the greenhouse effect is predicted to cause changes in precipitation patterns and evapotranspiration. According to climate change forecasts precipitation will shift to the winter half-year (OcCC/ProClim 2007). This shift results in elevated groundwater recharge, as in winter month evapotranspiration losses are less pronounced than in summer month. However, during the summer month and within gravel aquifers, less groundwater recharge will occur by the infiltration of river water due to more pronounced dry periods and reduced river discharges.

For processes of river–groundwater interaction, also physical changes as viscosity effects have to be considered. With rising temperature water is less viscose. This can result in elevated infiltration rates. Especially during heat periods infiltration rates can rise significantly (more than 50%, e.g., Braga et al. 2007).

Also artificial recharge and groundwater use will change during the different seasons. Reduced river water infiltration will affect riverine water supplies. This results in higher artificial recharge to meet water supply demands. Elevated river water temperatures in summer will partially ostracize the use of river water for cooling purposes. However, the demands for cooling purposes for industrial buildings rises and the pressure on lower temperate groundwater use will rise, too. An additional increase of groundwater temperatures will result in cases that the use of groundwater for cooling purposes is restricted. Thereby, the various zones within urban groundwater bodies will react in a different way (see below).

As long-term dry periods probably will increase (OcCC/ProClim 2007), deeper groundwater levels are expected in the winter half-year which can extend to springtime. In case long-term dry periods in summer repeatedly occur over several years, the groundwater head will not regenerate over years. This was observed after the very dry summer 2003. During this heat period, the use of river water in Switzerland was restricted and required special authorization. However, in the vicinity of infiltrating rivers, regeneration occurred rapidly. This again reconfirms the approach to divide groundwater bodies into zones (see below and Sect. 3.2) for which the groundwater renewal can be investigated. The areal distribution and the transient character of groundwater recharge for aerially infiltrating precipitation water for the region of Basel might alter due to climate change (Sect. 3.2). Also more frequent extreme precipitation events will have an influence on infiltration patterns and might lessen groundwater quality.

Additional qualitative aspects have to be considered. The amount of dissolved organic carbon (DOC) in drinking water can be used as a proxy that rising air temperatures increase the carbon turnover within the upper soil horizons. Infiltrating precipitation water then carries more DOC to the groundwater. While soils may dry up more pronounced than in the past decades, preferential flow of rains with increased intensity may carry infiltrating water fast to greater depths (Gerke et al. 2010).

Von Gunten et al. (1991) observed a redox zonation in aquifers, which is induced by seasonal variations in river water temperature and increased loads of organic matter due to wastewater discharge. The degradation of organic micropollutants and dissolved organic matter (DOM) strongly depends on the redox milieu and the temperature. This can influence the color of the groundwater. Also biological processes and activities are enhanced and consequently oxygen demand of groundwater. These processes will result in the decline of oxygen concentrations and reduced groundwater quality.

The longtime average of air temperature for the Basel area is 9.7°C as measured for the reference period 1961 to 1990 (meteorological station Basel Binningen). In the period from 1991 to 2009 air temperature increased to 10.8°C, illustrating that air temperature between the two periods has increased by more than 1°C. Average long-term groundwater temperatures in the region range between 12 and 16°C. This shows that elevated groundwater temperatures can not only be influenced by rising air temperatures but that also further anthropogenic influences have to be considered.

In case anthropogenic impacts on urban GWB are not reduced or even impaired, a long-term increase of groundwater temperatures will occur. However, we suggest that in urban areas, climate change effects will be secondary. More pronounced are anthropogenic impacts and changes of the use of aquifer resources. In case no measures will be taken against this development, groundwater temperatures in the Basel area will rise aerially up to 14 to 18°C. These changes can differ regionally in the vicinity of rivers, artificial recharge, and groundwater use. Figure 3.18 illustrates the temperature development and trend observed in the river Rhine and downstream of subsurface buildings. This example illustrates that the influence of heated subsurface buildings that reach the groundwater has a far higher impact on groundwater temperature compared to infiltrating river water with a long-term positive trend.

3.4.2 Effects of Predicted Climate Change on Groundwater Vulnerability

Butscher and Huggenberger (2009) presented examples for modeling future groundwater vulnerability in karst regions. The discussion is to some extent also valid for urban GWBs (Sect. 3.3). Results illustrated the present-day challenges and chances in conjunction with evolving water resources in a changing climate. Because groundwater vulnerability depends on local recharge conditions and subsurface properties, the modeled impact of climate change on groundwater vulnerability is likely to be site specific. The results further indicate a decrease in short-lived contaminants as a result of climate change. The impact of persistent contaminants, however, can only be determined if future climatic conditions at the site can be estimated with sufficient accuracy, because predicted summer heat waves and severe rainfall events will have opposite effects on groundwater vulnerability.

The anticipated effects of predicted climate change on the conduit flow vulnerability of a karst aquifer, represented by the dynamic vulnerability index DVI (see Sect. 3.3), are illustrated in Fig. 3.19. The simulation using the scenario "summer heat wave" results in generally shorter and less frequent periods with a high DVI as opposed to the simulation that was conducted with the observed (present-day) climate conditions from the simulation period. The effect is most obvious during

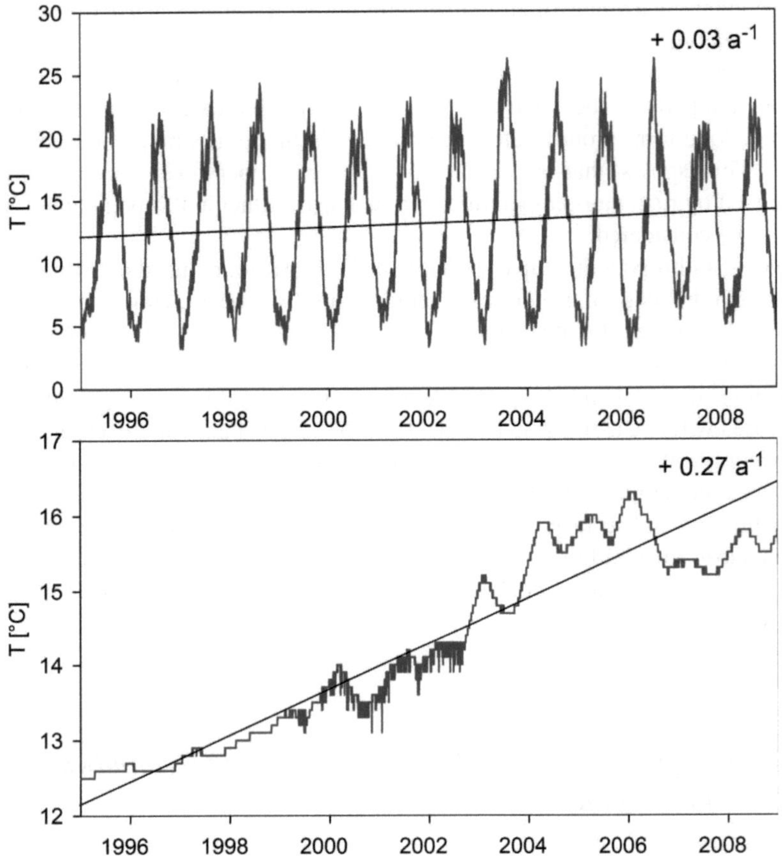

Fig. 3.18 Temperature development and trend observed in the river Rhine (*left*) and downstream of subsurface buildings (*right*)

spring and summer. The simulations indicate that the impact of short-lived contamination (e.g., fecal bacteria) is likely to decrease in the study area with the projected increased occurrence of hot and dry summers in the future. The results from the scenario "severe rainfall event" are in some respects astonishing. As expected, the conduit flow vulnerability rises dramatically directly after the event. In the long term, however, the conduit flow vulnerability slightly decreases as a result of enhanced recharge (also) to the diffuse flow system during the event. This recharge water is stored over quite some time, providing for an enhanced dilution of water from the conduit system during later events. Overall, this effect is advantageous for the drinking water supply: the long-term water quality is improved, while the short-term endangering of the water quality can be managed in most cases (e.g., by adapted raw water treatment such as chlorination or ozonation or by transient raw water rejection).

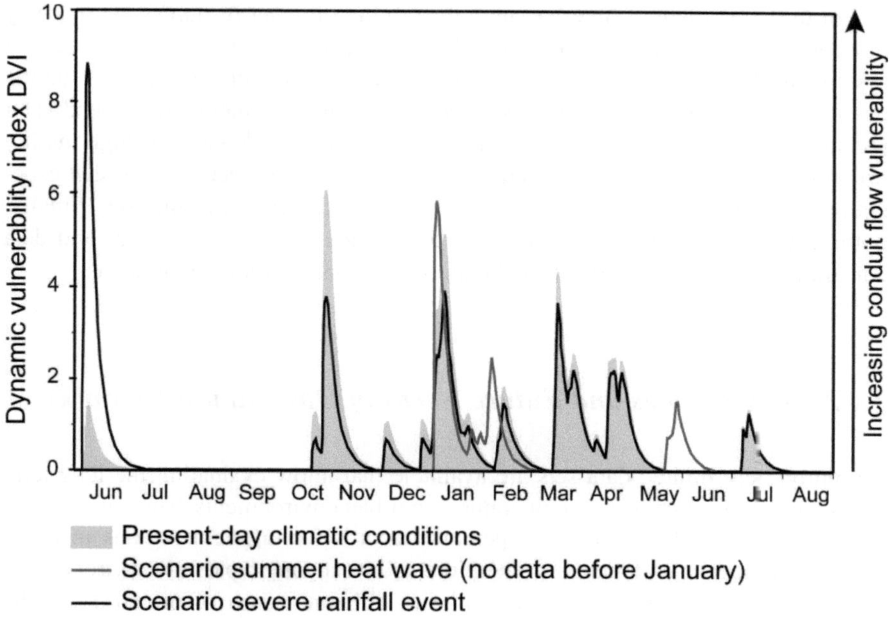

Fig. 3.19 Climate change and groundwater vulnerability. Simulated vulnerability index (VI) illustrating the expected effects of predicted climate change on "conduit flow vulnerability." Conduit flow vulnerability represents the sensitivity of the groundwater system to short-lived pollutants, e.g., fecal bacteria (Butscher and Huggenberger 2009)

The presented time variant modeling of vulnerability in Sect. 3.3 is a basic requirement for predicting the effects of climate change on groundwater resources. Both short-term variations and long-term trends can be evaluated using appropriate scenarios. Such modeling approaches are well suited for the spatial and temporal scales that are relevant for water suppliers. It can be adapted to local factors that will impact future groundwater quality, such as nutrient leaching in agricultural zones or mined areas. Data requirements are moderate and model handling is straightforward for professionals. Site specific knowledge of the evolving groundwater vulnerability in a changing climate will enable water suppliers and local authorities to take the necessary steps towards a sustainable management of the water resources. These steps include measures to preserve drinking water quality by tightened land use restrictions in the recharge areas or the installation of adequate water treatment systems.

As recharge can be considered as a key parameter for vulnerability evaluation (e.g., Aller et al. 1987), urban processes which modify groundwater recharge have to be included in any vulnerability assessment of given urban groundwater resources (see Sect. 3.3). Also local modifications of hydraulic conductivity of soil/subsoil formations by human activity have to be integrated in urban groundwater vulnerability assessments and mapping methods.

For the assessment of time-variant vulnerability of water resources, particularly the complex processes in river–groundwater interaction and the use of aquifer systems for heating or cooling have to be considered. This includes the evaluation of extreme flood and drought events (in context to climate change) on groundwater recharge processes and the consideration of an increased demand of high quality drinking water during and after major flood events. Experience showed that especially then great amounts of water are consumed for cleanup purposes. Finally, optimization techniques taking into account a wide range of objectives and data constraints can be tested on selected data sets of characteristic urban areas.

3.4.3 GWB Zones and Future Needs of Observation Networks

Currently, only limited data sets are available that allow evaluating the long-term development of groundwater temperatures in urban environments. The variations in daily, seasonal, and long-term temperature records for different regions in urban GWBs can show large deviations. Therefore, to investigate impacts of anthropogenic and climate change it is useful to delineate different zones of urban GWBs by analyzing historical groundwater temperature data.

Figure 3.20 shows how the urban GWB of Basel-Stadt was subdivided into zones according to the variation in temperature records and the expected dominating boundary conditions, which include the three rivers Rhine, Wiese, and Birs, the adjacent hills slopes Dinkelberg, Gempen, Bruderholz, and Tüllingen as well as the water production areas Wiese and Hardwald (see Sect. 3.2). The validity of the

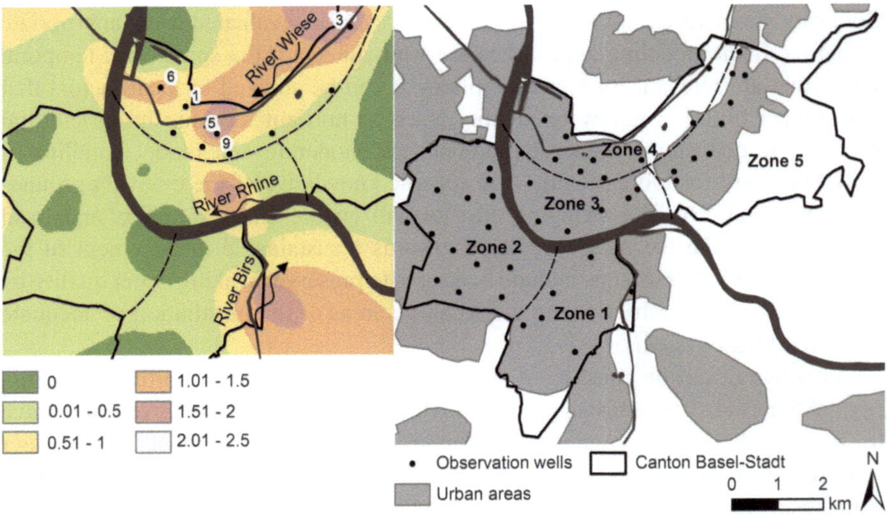

Fig. 3.20 Temperature variability illustrated for the year 2009 and delineation of zones for the GWB of the canton Basel-Stadt

delineated zones was confirmed by statistical analyses (SOM) of the temperature records (see Sect. 4.5). Subsequently, for the different zones further multivariate statistical analyses (PCA) were conducted to evaluate the influence of the various system borders individually for the regions of the different observation wells (see Sect. 4.5).

To capture the impacts of anthropogenic and climate change, in a next step, the setup of future observation networks can be optimized according to the results of the temperature data analysis in the various zones of urban GWBs. As long-term time series are indispensible for evaluating changes and making prediction on the long-term development of groundwater temperatures in urban environments, it is important to install new devices as soon as possible.

In Sect. 5.5, we present state-of-the-art observation wells for depth-oriented temperature monitoring that focus on capturing several spatiotemporally extremely heterogeneous processes, as (1) river–groundwater interaction effects on the thermal regime near the river; (2) temperature stratification downstream of thermal groundwater use; and (3) effects of construction parts reaching into aquifers and their downstream influence on groundwater resources. Further locations for new observation wells are suggested to capture the boundary conditions (see above) of the various zones of urban GWBs.

Often the spatiotemporal localization and the magnitude of anthropogenic and climate changes cannot be evaluated as there are not enough data and considerable uncertainties. To adequately detect and evaluate the influences of changes on urban groundwater resources, it is necessary to condition existing long-term data sets and to evaluate the influence of long-term annual, seasonal, and event driven developments of groundwater temperatures for different zones of urban GWBs and to assess the progression of groundwater temperatures for different depth of the aquifer within these zones. This includes the installation of additional monitoring devices to capture site-specific issues of climate and anthropogenic change, including supplementary monitoring stations for head, temperature, and oxygen measurements. The goal for the next couple of years is the implementation of robust instruments that allow evaluating and differentiate the various boundary fluxes to urban GWBs.

References

Affolter A, Gantenbein-Demarchi C, Huggenberger P, Lüthi T, Krapf T, Zoller R (2009) Verfahrensrichtlinie Trinkwasser (in Vernehmlassung)

Affolter A, Huggenberger P, Scheidler S, Epting J (2010) Adaptives Grundwassermanagement in urbanen Gebieten – Ansätze zur konkreten Umsetzung einer nachhaltigen Wasserressourcenbewirtschaftung. Grundwasser. doi:10.1007/s00767-010-0145-6

Aller L, Bennett T, Lehr JH, Petty RJ (1987) DRASTIC: a standardized system for evaluating ground water pollution potential using hydrogeological settings. U. S. Environmental Protection Agency, Oklahoma

Bates BC, Kundzewicz ZW, Wu S, Palutikof JP (2008) Climate change and water. Technical paper of the Intergovernmental Panel on Climate Change. Technical report, IPCC Secretariat, Geneva

Braga A, Horst M, Traver AG (2007) Temperature effects on the infiltration rate through an infiltration basin BMP. J Irrig Drain Eng 133(6):593–601

Brouyère S (2004) A quantitative point of view of the concept of vulnerability. In: Zwahlen F (ed) Vulnerability and risk mapping for the protection of carbonate (Karst) Aquifers. COST action 620, Final report, European Commission, Brüssel

Butscher C, Huggenberger P (2007) Implications for karst hydrology from 3D geological modeling using the aquifer base gradient approach. J Hydrol 342:184–198

Butscher C, Huggenberger P (2008) Intrinsic vulnerability assessment in karst areas: a numerical modeling approach. Water Resour Res 44:W03408

Butscher C, Huggenberger P (2009) Modeling the temporal variability of karst ground water vulnerability, with implications for climate change. Environ Sci Technol 43(6):1665–1669

Conversi A, Umani SF, Peluso T, Molinero JC, Santojanni A, Edwards M (2010) The Mediterranean Sea regime shift at the end of the 1980s, and intriguing parallelisms with other European basins. PLoS ONE 5(5):e10633. doi:10.1371/journal.pone.0010633

Doerfliger N, Jeannin P-Y, Zwahlen F (1999) Water vulnerability assessment in karst environments: a new method of defining protection areas using a multi-attribute approach and GIS tools (EPIK method). Environ Geol 39(2):165–176

Dyck S, Peschke G (1995) Grundlagen der Hydrologie, Berlin, pp 536

Eiswirth M, Hötzl H, Cronin A, Morris B, Veselič M, Bufler R, Burn S, Dillon P (2003) Assessing and improving sustainability of urban water resources and systems. RMZ Mater Geoenviron 50:117–120

Epting J, Huggenberger P, Rauber M (2008) Integrated methods and scenario development in urban groundwater management, and protection during tunnel road construction; a case study of urban hydrogeology in the city of Basel, Switzerland. Hydrogeol J 16:575–591

EU (2000) Richtlinie 2000 /60 /EG des Europäischen Parlaments und des Rates vom 23. Oktober 2000 zur Schaffung eines Ordnungsrahmens für Maßnahmen der Gemeinschaft im Bereich der Wasserpolitik

Fatta D, Naoum D, Loizidou M (2002) Integrated environmental monitoring and simulation system for use as a management decision support tool in urban areas. J Environ Manage 64:333–343

Fogg G, LaBolle E, Weissmann G (1999) Groundwater vulnerability assessment: hydrogeologic perspective and example from Salinas Valley, California. In: Corwin D, Loague K, Ellsworth T (eds) Assessment of non-point source pollution in the Vadose zone, Geophysical Monograph No 108. American Geophysical Union, pp 45–61

Gerke HH, Germann P, Nieber J (2010) Preferential and unstable flow: from the pore to the catchment scale. Vadose Zone J 9:207–212. doi:10.2136/vzj2010.0059

Hari RE, Livingstone DM, Siber R, Burkhardt-Holm P, Guettinger H (2006) Consequences of climatic change for water temperature and brown trout populations in Alpine rivers and streams. Glob Chang Biol 12(1):10–26

Hötzl H (1996) Grundwasserschutz in Karstgebieten, Grundwasser 1/96 (1996):5–11

INTERREG III A-Projekt MoNit (2006) "Modellierung der Grundwasserbelastung durch Nitrat im Oberrheingraben" Landesanstalt für Umwelt, Messungen und Naturschutz Baden-Württemberg

Jeannin P-Y, Cornaton F, Zwahlen F, Perrochet P (2001) VULK: a tool for intrinsic vulnerability assessment and validation. In: 7th Conference on limestone hydrology and fissured media, Besanc, 20–22 Sept 2001. Sci Technol Environ Mem HS 13:185–190

Liedl R, Sauter M, Hückinghaus D, Clemens T, Teutsch G (2003) Simulation of the development of karst aquifers using a coupled continuum pipe flow model. Water Resour Res 39(3):1057. doi:10.1029/2001WR001206

National Research Council (1993) Groundwater vulnerability assessment, contamination potential under conditions of uncertainty. Committee on Techniques for Assessing Ground Water Vulnerability, Water Science and Technology Board, Commission on Geosciences Environment and Resources, National Academy Press, Washington DC, pp 224

National Research Council (2004) Adaptive management for water resources project planning, 138 S. National Academies Press, Washington

OcCC/ProClim (ed) (2007) Climate change and Switzerland 2050. Expected impacts on environment, society and economy. OcCC/ProClim, Bern

Pahl-Wostl C, Möltgen J, Sendzimir J, Kabat P (2005) New methods for adaptive water management under uncertainty–the NeWater project. In: Paper accepted for the EWRA Conference 2005, Menton, France, September 2005

Pronk M, Goldscheider N, Zopfi J (2007) Particle-size Distribution As Indicator for Fecal Bacteria Contamination of Drinking Water from Karst Springs. Environmental Science & Technology 41(24):8400–8405

Rauch W (2009) Anwendung des HACCP Konzepts (Hazard Analysis and Critical Control Points) zum Schutz eines Trinkwasserbrunnens. GWF 07–08

Schirmer M, Strauch G, Schirmer K, Reinstorf F (2007) Urbane Hydrogeologie – Herausforderungen für Forschung und Praxis.-. Grundwasser. doi:10.1007/s00767-007-0034-9

Stanford JA, Ward JV (1993) An ecosystem perspective of alluvial rivers: connectivity and the hyporheic zone. J N Am Benthol Soc 12(1):48–60

von Gunten HR, Karametaxas G, Krähenbühl U, Kuslys M, Giovanoli R, Hoehn E Kei R (1991) Seasonal biogeochemical cycles in riverborne groundwater. Geochim Cosmochim Acta 55(12):3597–3609

Vrba J, Zoporozec A (1994) Guidebook on mapping groundwater vulnerability. In: Vrba J, Zoporozec A (eds) International contributions to hydrogeology (IAH), vol 16. IAH, Hannover, p 131

Water for health (2010) who guidelines for drinking-water quality. http://www.who.int/water_sanitation_health/dwq/guidelines/en/

Chapter 4
Methods

Peter Huggenberger, Jannis Epting, Annette Affolter, Horst Dresmann, Ralph Kirchhofer, Edi Meier, Rebecca M. Page, Christian Regli, Jelena Simovic Rota, and Stefan Wiesmeier

A comprehensive knowledge of the subsurface structure and composition (geomechanical properties) together with flow processes is a prerequisite for developing innovative concepts for resource protection and management in urban environments. With increasing development pressure, the integration of this knowledge in planning procedures becomes essential for a sustainable and efficient use of the urban subsurface.

In the last 10 years, for the Basel area, a comprehensive geological database was set up. Additionally, numerous ongoing projects involved the development of new methods that allow testing rules for urban subsurface planning and management. Several investigation areas have been equipped with extensive groundwater monitoring systems. At the same time, high-resolution geological and hydrogeological models were set up and calibrated with long-term datasets. The combination of the various investigation methods is the basis for hypothesis testing.

This chapter summarizes some important methods applied individually or in combination for the different case studies. First the role of geological and hydrogeological data is discussed and an efficient as well as comprehensive database is presented. Subsequent sections introduce the concept of adaptive resource management as well as concepts for monitoring and applied modeling approaches. In addition, selected hydrogeophysical investigation methods and methods that allow characterizing aquifer heterogeneity and to statistically analyze complex data are presented.

4.1 Data Mining

Horst Dresmann, Ralph Kirchhofer, and Stefan Wiesmeier

In urban areas, large amounts of geological, hydrological and geotechnical data are available. However, they often are not systematically organized and are spread in archives of numerous institutions or companies. As long as such data are not

P. Huggenberger and J. Epting (eds.), *Urban Geology: Process-Oriented Concepts for Adaptive and Integrated Resource Management*, DOI 10.1007/978-3-0348-0135-0_4,

compiled systematically they essentially are of no value. The expanding use geodatabases and of Geographic Information Systems (GIS) in environmental sciences and the growing acceptance of a need for quality management of data have necessitated reappraisal of data management needs.

The development of concepts for data mining and access tools on the basis of open interoperability standards improve the efficiency of using the existing information for environmental projects and research issues. Such concepts should deal with the following questions that generally arise in the context of urban geology and hydrogeology:

1. What data exist, are the available data accessible and sufficient for geological, hydrological and geotechnical investigations?
2. How can data acquisition and exchange be optimized?
3. What tools allow quick access to the data?

Some fundamental requirements for data management tools and further applications include the possibility to flexible design data queries, perform data analysis and simple statistics, analyze and visualize spatial information and to export customized data formats for geologic and hydrogeological models and software.

Subsequently, for the specific geological, hydrological and geotechnical investigations, the data management tools allow formulating questions as: What questions can be answered with the existing data or what additional data can improve the investigations and predictions?

In most urban areas large amounts of data are available but spread at different institutions or companies. These data often are difficult to access and are not systematically organized. The most important constraints in this respect are:

1. Economical: some institutes have to charge a fee to cover costs for the data collection.
2. Political: conflicting interests among data collectors and/or data users can hamper data access.
3. Data formats: data rarely exist in harmonized digital formats. Furthermore, some data are only available on paper.
4. Fractionized databases: often data exist in many databases with different data owners, different data formats and dissemination policies.
5. Transboundary barriers: there are problems in harmonizing and exchanging data in international river basins.

Many geological, hydrological and geotechnical questions in urban environments can be addressed more efficiently if an accurate database of the subsurface properties exists. This efficiency also results from investing more time in the solution process in general than in the effort of barely collecting data for documentation. Hence, in this section, we present a possible setup of a geological database that was started in the late 1980s and which was continuously developed.

4.1.1 Data Mining with GeoData

Due to the fact that localization and assemblance of geological data are time consuming, a first version of a *Geo*logical *Data*base (GeoData) system was realized by Noack (1993) for specific projects (i.e., earthquake microzonation) in the Canton Basel Stadt. Kirchhofer (2006) enhanced GeoData and included an application for automatically classifying sedimentary structures of borehole descriptions. The development of GeoData was accompanied by the environmental agencies of Basel-Stadt and Basel-Landschaft (Switzerland). This interface allows the direct provision of customized data for numerous cantonal projects. Furthermore, GeoData is used as a service for the public domain by consulting hydrogeologists and geologists, urban planners and water supply engineers as well as environmental agencies. Altogether, GeoData is a unique data source suitable for empirical studies and hypothesis testing in the field of quantitative information on urban geological and hydrogeological questions (Fig. 4.1).

GeoData comprises systematic data collection, analysis of borehole descriptions and the assessment of metadata from geological, hydrological and geotechnical reports. It includes more than 10,000 borehole descriptions and metadata from more than 3,700 reports (Fig. 4.2). Tailored database applications from GeoData are used for the calculation of the 3D aquifer base, the evaluation of aquifer parameters and for lithofacies-based interpretations of borehole descriptions (see below).

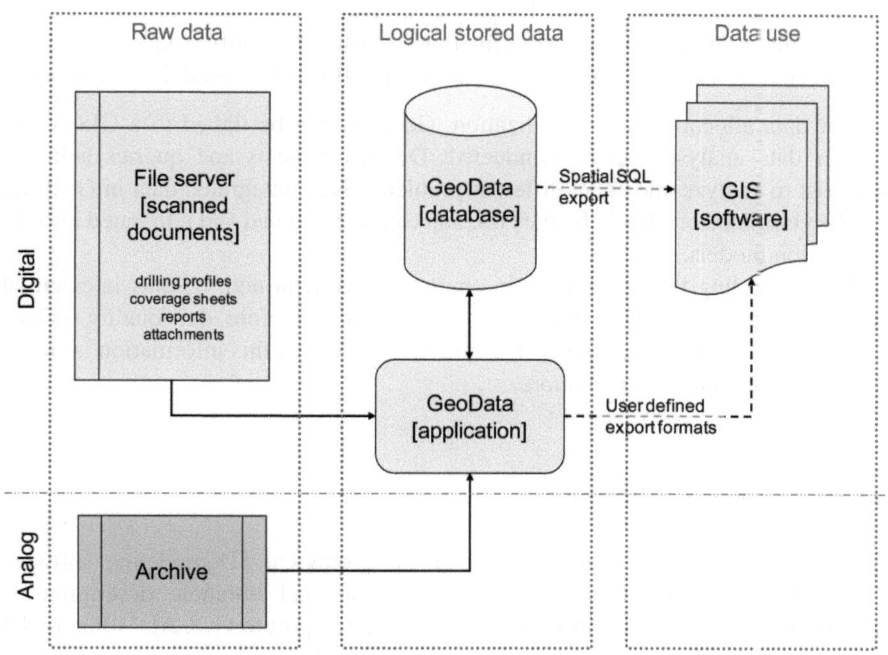

Fig. 4.1 General structure of GeoData

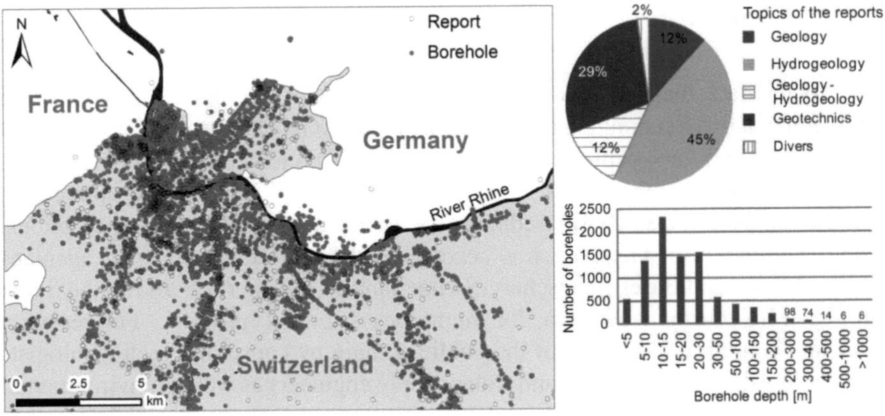

Fig. 4.2 *Left*: Overview illustrating the density of regional borehole data and reports. *Top right*: Distribution of reports according to main investigation subject. *Bottom right*: Number of boreholes in various depth classes

Figure 4.1 shows the structure of GeoData embedded in the software environment. Main elements of GeoData are:

1. Full control of data input, change and deletion.
2. Advanced search functions and differentiated queries.
3. Exports for different geological and hydrological software.
4. Automated report generation for public services.
5. Preliminary data analysis and evaluation of different quality data.
6. Direct access to a GIS based on the object-relational data model of GeoData.

For data allocation and visualization, GeoData can be linked to a GIS where further data analysis can be conducted. Defined projects and queries help the scientist to analyze and solve different problems. New integrated data in GeoData can be visualized on-the-fly by the GIS, and can be evaluated and integrated into the simulation models.

Understanding the uncertainty in environmental data and systems is essential for making robust and wise management decisions. Therefore, data quality is characterized in terms of uncertainty. For borehole data, e.g., this information is stored as part of the data documentation.

4.1.1.1 Geological Horizons

GeoData includes an export function to the geological 3D modeling software GOCAD®. Thereby information from the geological borehole description is transformed into the well format necessary for the import to GOCAD®. Figure 4.3 shows the base unconsolidated rock for the Basel area deriving from the data of GeoData and interpolated in GOCAD®. Additional datasets from DEM as well as

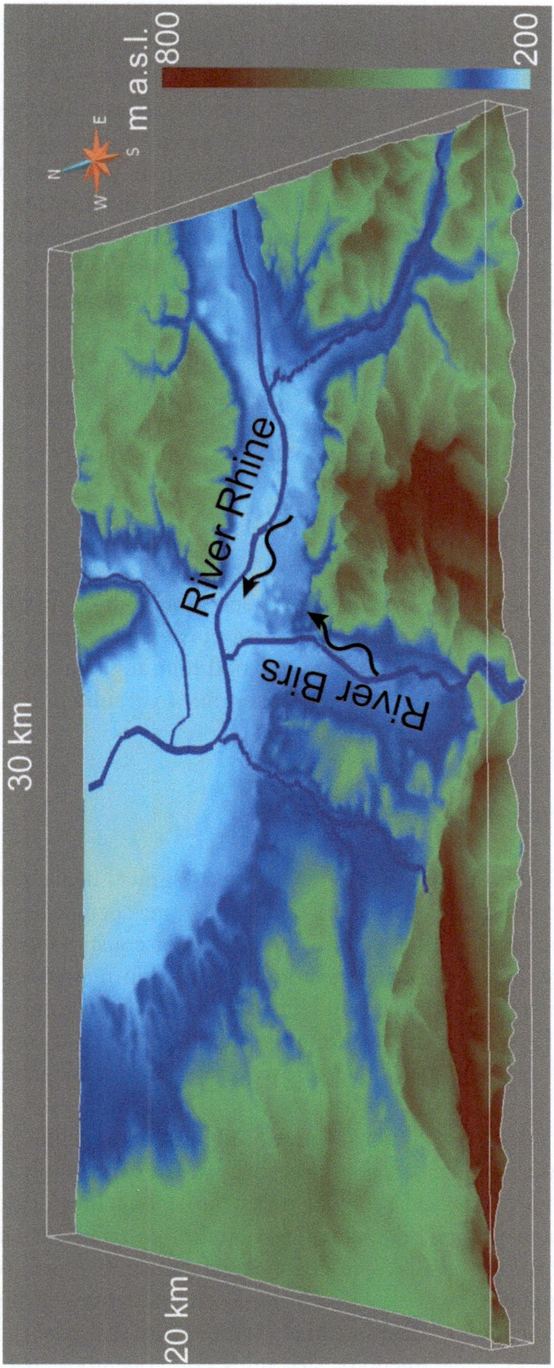

Fig. 4.3 Base unconsolidated rock of the Basel area

further geological and geomorphological interpretation approaches served as the basis for an advanced 3D horizon modeling.

In a next step a characterization for the quality of the modeled horizon was established. This was necessary to address the uncertainty of derived information from the modeled horizon for engineers and model applications. Therefore, the various input data were classified into more or less resilient data. Information on the base unconsolidated rock from borehole descriptions and from bedrock outcrops are considered as rather reliable. Those, which are derived from geological and geomorphological interpretation approaches, are more subjective and hence considered as less reliable compared to the preceding ones.

Exports from GeoData to model the deeper geological horizons in GOCAD® allow to analyze thickness distributions of geological units. Also the identification and characterization of faults within the borehole descriptions are essential for modeling faults and tor interpreting thickness distributions of geological units across faults.

4.1.1.2 Aquifer Parameters

Distribution of horizontal conductivity zones for hydrogeologic models can be derived from different type and quality datasets available from GeoData. Furthermore, additional information from reports outlining regional geological questions, including pumping test data, is used to determine hydraulic parameter distribution.

4.1.1.3 Interpretation of Lithofacies

Sedimentological borehole descriptions of coarse gravel deposits provide information on composition and texture of deposits. More specifically, some details of grain-size categories, sorting, composition of major constituents and proportion of each grain-size fraction can be determined and information obtained on color, chemical precipitation, thickness of a deposit, and its transition through the underlying layer. However, the information content depends on the quality of the geological descriptions, which are to some degree subjective and can vary considerably.

Based on borehole descriptions, defined sedimentary structure patterns and on the interpretation method for borehole descriptions, GeoData allows analysis of sedimentary structure types and provides datasets of point information with arbitrary separation distances along boreholes. The point information includes space data (*x*-, *y*-, *z*-coordinates), probabilities of sedimentary structure types (probability that a borehole descriptions layer description represents a defined sedimentary structure type) and an indication of the most likely sedimentary structure type (Fig. 4.4). The spatial distribution of sedimentary structures consequently can be integrated into geologic and hydrogeologic models which will be discussed in Sect. 4.4.

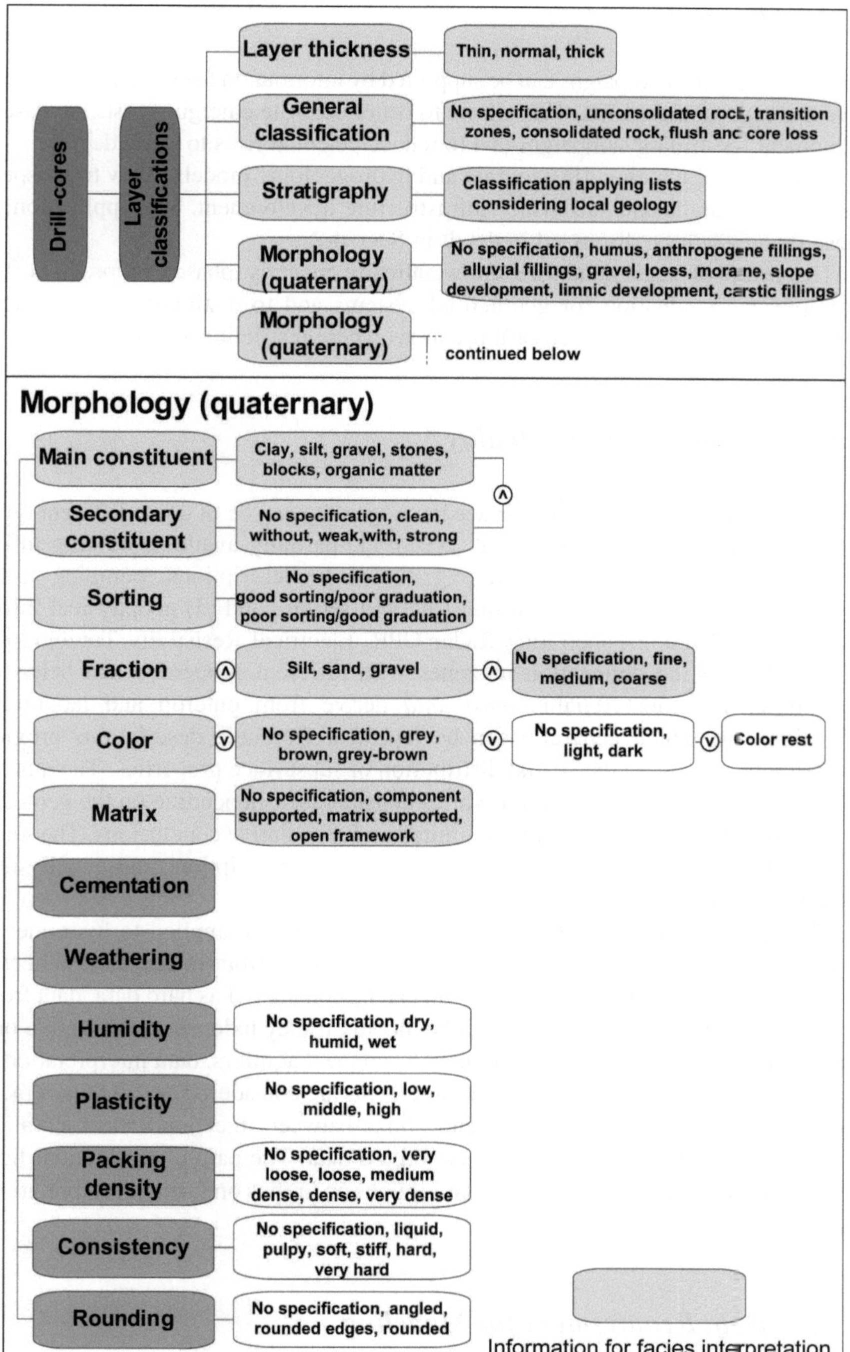

Fig. 4.4 Borehole input mask for GeoData (V: or; Λ: and). The information used for sedimentary structure types interpretation is *highlighted*

4.1.1.4 Prognosis

Upcoming drilling campaigns can be supported by information from existing borehole data and report information. This allows to better estimate emerging costs, necessary equipment for drilling campaigns and to reduce potential risks to some degree.

Additionally, existing 3D geologic and hydrogeologic models allow to prospect drilling campaigns and subsurface infrastructure development. The application of these models will be discussed in detail in Sect. 4.2.

The setup of such databases allow, already in early phases of projects, for example, the evaluation for geothermal systems and to distinguish areas where these systems can be installed or those areas where restrictions are necessary.

4.1.2 Evaluating Data Quality

The various information on subsurface heterogeneity can be of quite different type, quantity and quality (cf. Regli et al. 2002): (1) partially available outcrop information; (2) borehole information (e.g., borehole descriptions, pumping tests) providing only a limited view of the structural setting; and (3) geophysical information (e.g., Ground Penetrating Radar GPR, Electrical Resistivity Tomography ERT) allowing the delineation of zones with different properties and behavior over time. The most reliable *hard data* derive from outcrop and laboratory investigations, which generally can be repeated. Borehole descriptions provide limited information on the spatial distribution of subsurface properties. The quality of individual borehole descriptions varies considerably, depending on the geotechnical approach used, thus permitting limited and speculative conclusions. The same is true of geophysical data. Consequently, borehole descriptions and geophysical data are regarded as *soft data.*

The designation of "hard" and "soft" data can also be applied to hydrometric data deriving from hydraulic measurements as well as from tracer tests. Whereas hydraulic measurements of the river head can be considered as hard data, data from groundwater observation wells are hard data only if they independently sample one aquifer. In the case where observation wells connect aquifers, data interpretation is not distinct and consequently such measurements are considered as soft data. Tracer tests can undoubtedly confirm hydraulic links between injection and observation locations; this information corresponds to hard data. The path of preferential flow between these locations, however, is ambiguous and relies on further interpretation, resulting in soft data.

4.1.3 Data Requirement for Modeling

Sufficient data availability and accessibility is a key factor for successful monitoring and modeling. Existing data are needed to characterize and assess the status

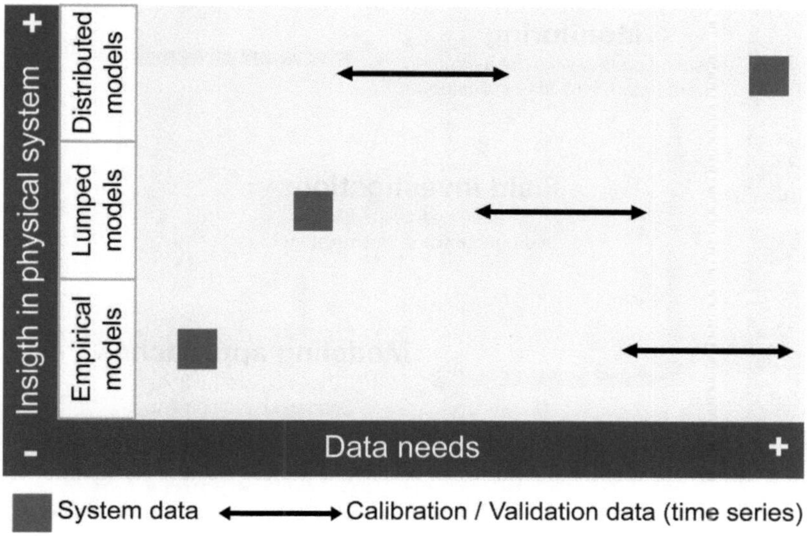

Fig. 4.5 Data requirements for different modeling approaches

of the urban subsurface and water bodies and new data are needed to verify or to follow the development of these. Often the question arises if there is enough data to support modeling.

Figure 4.5 illustrates that different model types require different amounts of data (Harmoni-CA Synthesis Report G-09). Simple empirical models that can be used for screening purposes only require relatively few input data, but need extensive calibration and validation datasets to obtain good model performance. The opposite is the case with a more comprehensive but complex distributed model, required in more complex management situations. These models need much more input data to perform correctly.

4.2 Elements for Adaptive Resource Management

Jannis Epting, Peter Huggenberger, Horst Dresmann, and Annette Affolter

The concepts and the basic principles for adaptive subsurface and water resource management have been introduced in Sect. 3.1. They include the identification of the current profiles of environmental systems and risks for the relevant scales. The approaches also include the formulation of specific targets that leads to the achievement of the overall development goals for the subsurface and water resources.

The basic elements for adaptive resource management include: (1) the setup of adequate observation networks for monitoring; (2) the selection of appropriate

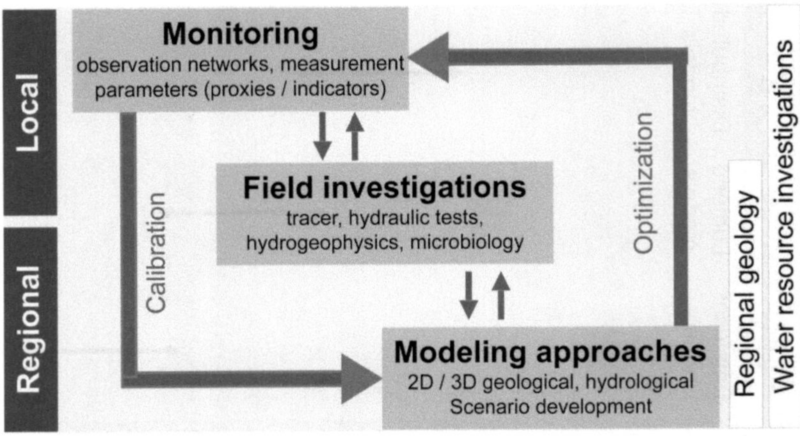

Fig. 4.6 Iterative adjustment of elements for adaptive subsurface resource management

modeling tools; as well as (3) the definition and realization of specific field measurements to close existing knowledge gaps.

In this section we discuss some general thoughts concerning the optimal design of observation networks and the appropriate selection of measurement parameters. Further we highlight the role of field investigations, whereas hydrogeophysical measurements are discussed in Sect. 4.3. Finally, we illustrated how to select suitable modeling approaches for testing hypotheses and to answer the different environmental questions.

As a result, we show that the solution process for environmental questions is an iterative procedure of consecutive data acquisition deriving from monitoring specific field investigations and experiments as well as the interpretation with the aid of geological and hydrogeological modeling approaches (Fig. 4.6).

4.2.1 Monitoring

The installation of observation networks for monitoring hydraulic, physical and chemical parameters is the basis for all geological, hydrogeological and geotechnical investigations and the resulting data are prerequisite for the setup and calibration of geological and hydrogeological models.

Answers to the following questions help to optimize the setup of observation networks:

1. Where are data gaps, which parameters are reasonable to be measured, which data can be used as indicators and/or proxies for specific processes (see Sect. 3.3)?
2. How can measurements on different temporal and spatial scales optimally be combined?

3. How can new boundary conditions be specified?
4. How can observation networks be optimized, primarily with respect to minimizing the uncertainty in the quantity to be assessed for given costs and other constraints, but also with respect to installation, communication, automatic data storage and data retrieval?

Observation networks enable to monitor the current status of environmental systems. The data from observation networks then can be incorporated into subsurface and groundwater management concepts to qualitatively and quantitatively safeguard subsurface resources and groundwater bodies. This also includes the long-term improvement of groundwater quality within the catchment zones of water supplies and downstream of thermal groundwater use.

An example for capturing the transient character of river–groundwater interaction and the setup of observation systems for depth-oriented groundwater measurements near in- or exfiltrating river segments is presented in Sect. 3.2. Such groundwater observation networks enable to monitor different parameters and to derive proxies.

In Sect. 5.5 state-of-the-art observation wells for depth-oriented temperature monitoring are presented that focus on capturing several spatiotemporally extremely heterogeneous processes, such as (1) river–groundwater interaction effects on the thermal regime near the river; (2) temperature stratification downstream of thermal groundwater use; and (3) effects of construction parts reaching into aquifers and their downstream influence on groundwater resources.

We exemplify the setup of an appropriate observation system for a complex groundwater body that is located in an area with contaminated sites and where important drinking water production takes place. The drinking water production is protected by the infiltration of filtrated Rhine water, creating a hydraulic gradient towards the known contaminated sites and towards infiltrating surface water from the river Rhine. In Sect. 5.3 we analyze the development of groundwater pollution during the last decades in this heavily industrialized groundwater protection area. This includes the illustration of long-term changes to the groundwater body due to changed hydraulic boundary conditions.

An observation system with 121 observations wells was set up (Fig. 4.7) to monitor and document the evolution of the contaminations in the various areas as well as to define the boundary conditions and to calibrate a 3D numerical groundwater flow model. As in this region two aquifers are of importance, observation wells for depth-oriented monitoring were developed which allow to sample the lower karst aquifer.

4.2.2 *Field Investigations and Experiments*

Next to monitoring data, the various conceptual and numerical models require validation by means of specific field investigations and experiments. These

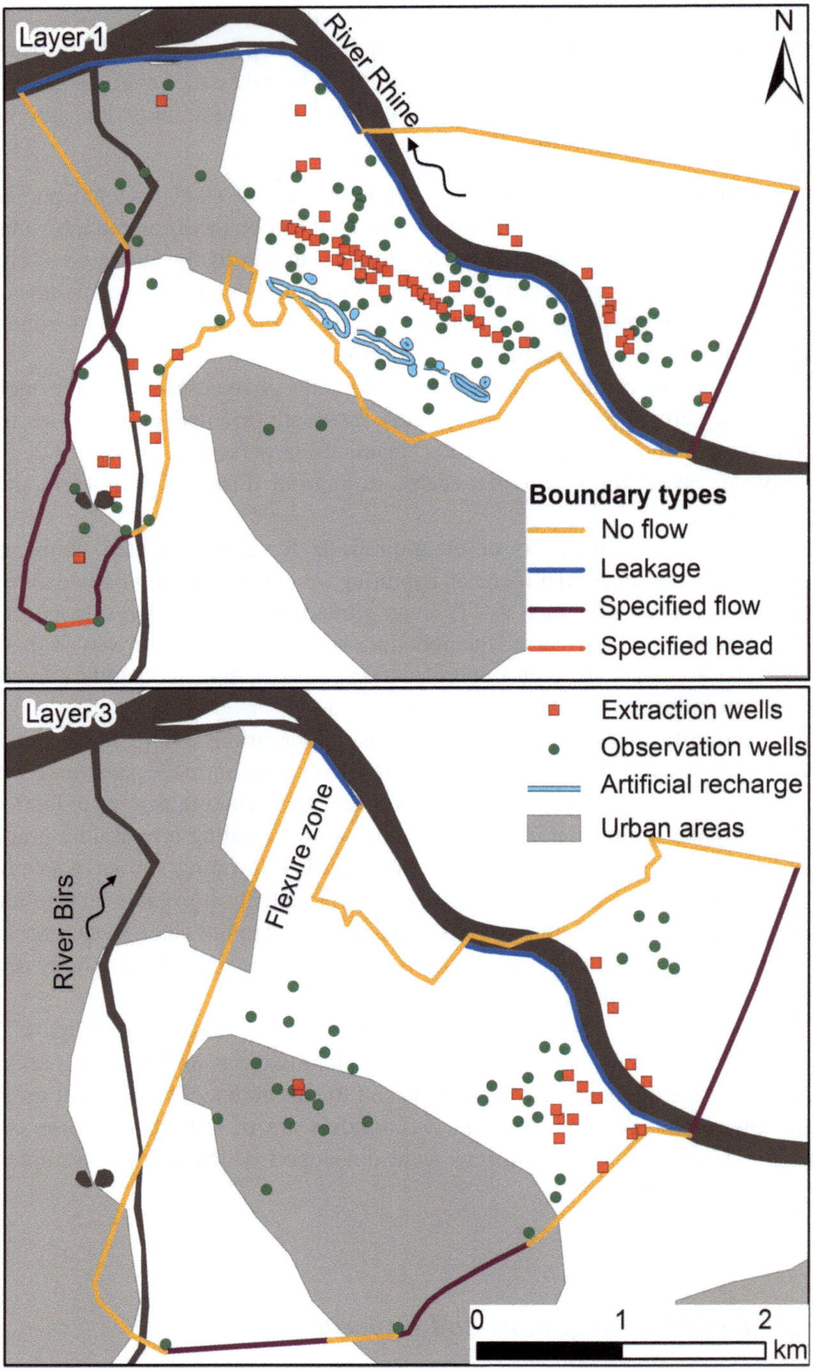

Fig. 4.7 Setup of an observation system for a complex GWB in the Basel area and definition of boundary conditions for a 3D hydrogeological model. *Top*: Upper gravel aquifer; *Bottom*: Lower Muschelkalk aquifer

approaches allow to test previously formulated questions and hypotheses and to achieve defined goals for GWBs (Pahl-Wostl 2006). In addition, such investigations and experiments are important elements of adaptive groundwater management which involves a structured iterative process to approach complex groundwater systems. Further, this procedure is prerequisite for subsequently defining scenarios for different project phases. When conducting field investigations or experiments, the question arises from what kind of approach can basic information be derived that allows resilient decisions?

Field investigations include among others: (1) tracer and hydraulic tests; (2) hydrogeophysical investigations as highlighted in Sect. 4.3; and (3) event-oriented hydrochemical and microbiological sampling. The realization of specific field investigations enable studying the transient character of the relevant processes and boundary conditions; as well as extrapolating 1D information of monitoring results into 2D zones of enhanced or reduced process activities (see Sect. 4.3). Some experiments, to constrain density-driven flow modeling approaches, are presented in Sect. 5.4.

4.2.3 *Modeling*

Geological and hydrogeological models of the subsurface help to better understand natural systems and processes and allow assessing their reaction to technical interferences or to changed environmental situations.

Our modeling experience derives from a large number of projects in the region of Basel. In this section we distinguish between geological and hydrogeological models. In the process of geological modeling, on the one hand, the aim is to integrate different kinds of geological data (see below) and to derive structural 2D and 3D maps, which further can be used for properties modeling of solids. The hydrogeological models, on the other hand, use the input data for the model geometries that are derived from the geological models. They are mainly used to model groundwater flow and transport regimes, hydrogeologic and hydrologic processes and to evaluate the changes due to infrastructure development.

Further, these types of models are the core elements for planning and the basis for defining management rules for the different, often conflicting interests concerning the use of the urban subsurface. For many urban geological and hydrogeological issues, models can be used as a basis for discussion.

Geological and hydrogeological models often are an integral part of studies. They help to structure the processes of both sampling and evaluation. Often these models need to be refined (also by specific field investigations or experiments, see above) to the relevant scales and adjusted to the study objectives and adequate planning procedures.

The development and refinement of methods for spatial 3D planning and process understanding of the subsurface significantly assist in answering geological and hydrogeological questions. Thereby, the use of numerical simulation models is suited during several project phases. On one hand, such models can be incorporated to define the current status of the subsurface and groundwater regimes and, on the other hand, they can be used to prospect future changes to the subsurface.

In this section we do not make the attempt to give an all-encompassing overview on modeling techniques. An overview of the various models and their applications is outlined extensively in the literature (e.g., http://water.usgs.gov/software/lists/groundwater/). Therefore, we present those modeling approaches which we applied for the specific management approaches in the region of Basel. We have developed several high-resolution geological and hydrogeological 3D models, which are applied in the context of urban geology and hydrogeology in the Basel area.

4.2.3.1 Regional Geological 3D Model of the Basel Area

The regional geological 3D model includes all relevant geological horizons in the Basel area and can be refined to local geological 3D models covering the planning scale of subprojects (Fig. 4.8). The main elements of the geological 3D models are:

1. Geological information from borehole descriptions, geological mapping and outcrops as well as derived from geophysical surveys.

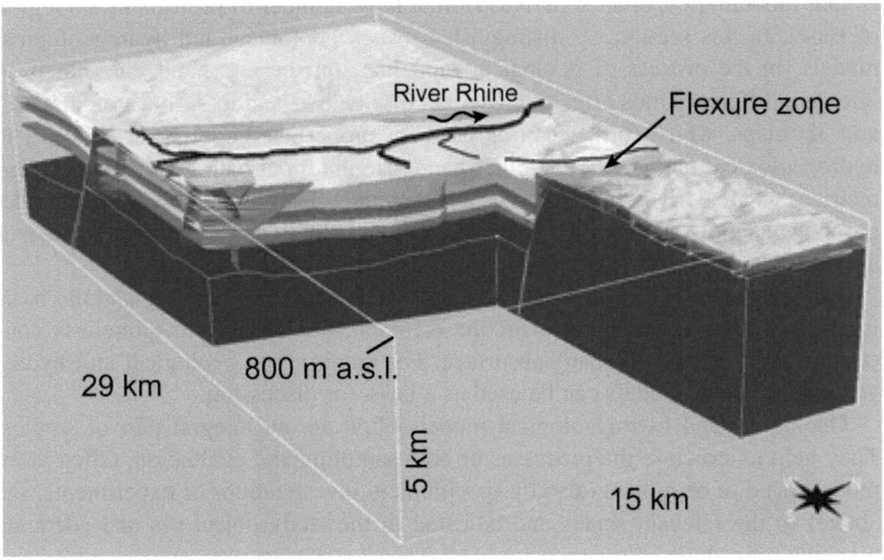

Fig. 4.8 A part of the regional geological 3D model of the Basel area showing the geological horizons (cf. Table 2.1) as well as selected faults and tectonic features

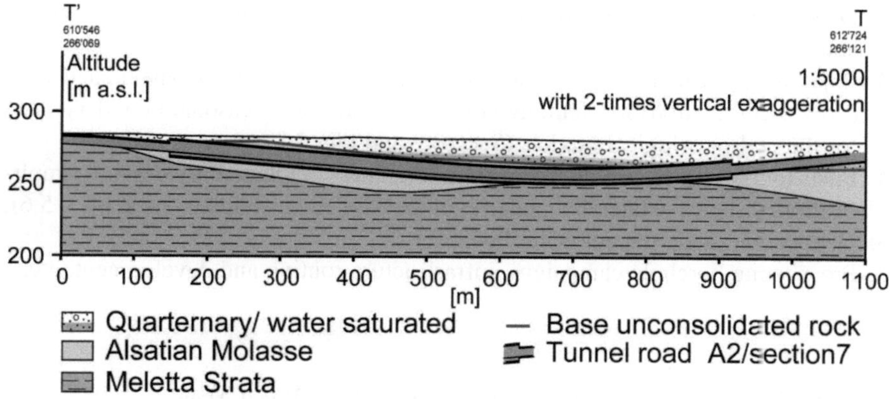

Fig. 4.9 Geological profile as basic information for a tunnel construction in the Basel area

2. Construction of geological horizons or layers, geological solids.
3. Representation of fault systems and tectonic features.

Subsequently, the geological 3D models can be transformed to coupled 3D groundwater flow, transport and heat transport as well as geomechanical models to investigate specific processes (influence of artificial subsurface bodies on groundwater flow processes, mechanical risks during subsurface construction work, effects of thermal groundwater use, etc.).

The integration of knowledge from geological 3D structures, geomechanical properties and subsurface processes can be used to address specific problems related to urban planning. As subsurface parts of buildings and technical infrastructures can be integrated into the geological 3D models, model-based predictions of impacts of urban development projects provide an important basis for decision making.

Figure 4.9 illustrates an application example where we derived basic geological information from the geological 3D model. The figure shows the progression of a projected tunnel road within the various geological formations. Especially, the interfaces between the single geological formations and transition into the gravel deposits represent locations where hazards might occur during the construction period of the tunnel road.

4.2.3.2 The Base Unconsolidated Rock of the Basel Area

All geological and hydrogeological 3D models include the information of the base unconsolidated rock of the Basel area which can be used for several applications:

1. Layer uncovered from Quaternary deposits, which is included into the local and regional 3D geologic models (Sect. 4.1 and Fig. 4.3).

2. Definition of the lower boundary for the hydrogeological models of gravel aquifers and water bodies.
3. Basis for the calculation of catchment areas for the base unconsolidated rock (flow accumulation) for defining boundary conditions (seasonal boundary flux) for the hydrogeological models (Sect. 3.2 and Fig. 3.8).
4. Basis for the earthquake microzonation (thickness of Quaternary deposits/amplification, distance to lateral boundaries of the base unconsolidated rock, Sect. 5.6).
5. Delineation of significant water bodies.
6. Prospecting borehole campaigns, infrastructure routing and development, etc.

4.2.3.3 Hydrogeological 2D and 3D Models of the Basel Area

With the various hydrogeological 2D and 3D models of the Basel area (Fig. 4.10) we mainly focus on water resource management and protection (quality assessment) for drinking and industrial water supply as well as during subsurface infrastructure development. Another focus is investigations of contaminant and heat transport as well as on the understanding of specific processes such as river–groundwater interaction or thermal stratification within aquifers.

Hydrogeological models are able to simultaneously include hydrological and hydrogeological as well as operational data which allow assessing the related

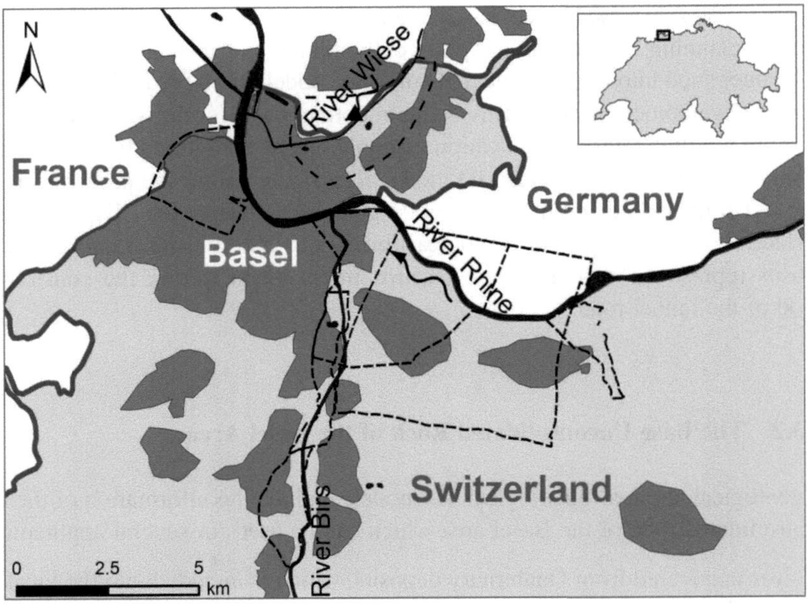

Fig. 4.10 Hydrogeological models of the Basel area

groundwater flow regimes. The main elements of the hydrogeological 2D and 3D models of the Basel area are:

1. Model geometries derived from the base unconsolidated rock and the geological models (see above).
2. Assignment of aquifer properties (calibrated, from aquifer tests, etc.), including the consideration of subsurface heterogeneity of fluvial deposits and specific karst phenomena in the deeper aquifers (conduits, zones with variable permeability, weathered zones, see Sects. 4.1 and 4.4).
3. Definition of boundary conditions (natural and operational).
4. Steady-state and transient flow and transport modeling.

The calibrated hydrogeological models allow evaluating inflow and capture zones to groundwater extraction wells (definition of protection zones, Fig. 4.11), water budgets through model boundaries and defined zones of interest as well as the development of scenarios (see below). They allow the optimization of quantitative and qualitative groundwater use as well as the evaluation of hazardous impacts, creeping contaminations from brown fields and nitrate on drinking water quality. Furthermore, modeling allows to optimize monitoring networks and field investigations and vice versa (Fig. 4.6).

Additionally, hydrogeological models can be used for the investigation of specific processes, as river–groundwater interaction, density coupled flow as well as contaminant and heat transport (see Sects. 5.4 and 5.5). The nature of calibration is not to model the processes in details but to model the bulk behavior of the

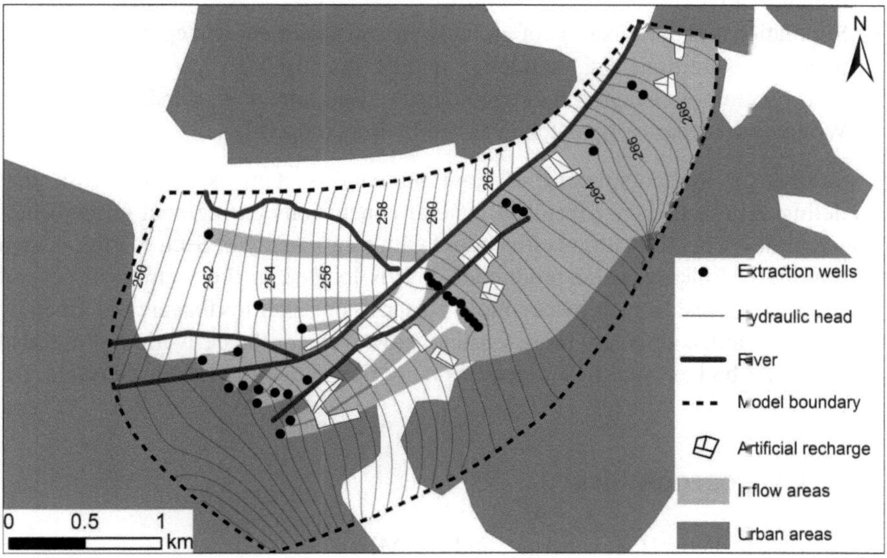

Fig. 4.11 Modeled inflow areas to the groundwater production wells in the Langen Erlen, Basel. Results with low hydrological (low river stage and low boundary fluxes) and average operational boundary conditions are shown. The approach allows to delineate the area of potential influence and to define target-oriented protection measures within the inflow areas of the wells

physical system (patterns of parameters, in/out flows, etc.) which gives one possible description of the aquifer.

4.2.3.4 Numerical Modeling and Scenario Development

Scenarios represent system states and event sequences and serve to acquire and illustrate a representative selection of possible dispositions and process sequences. Scenario development also involves the simplification and restriction of essential boundary conditions that affect the system. With the calibrated modeling tools, scenarios can be developed which facilitate the evaluation of system sensitivities, allowing the investigation of certain parameters and boundary conditions. Physical-based models describing the subsurface and groundwater systems can be the basis for optimization.

With the development of scenarios, possible impacts of subsurface constructions, water engineering measures and from hydrogeologic extreme events on groundwater regimes and groundwater usage can be acquired. Like this, the corresponding endangerment and risk assessments can be conducted. Scenarios can be assigned to four groups:

1. Evaluation and comparison of subsurface infrastructure development and routing as well as water engineering measures, with respect to feasibility and impact on GWB during construction and after completion (adapted technical boundary conditions, e.g., effects of water engineering measures, evaluation of engineering alternatives).
2. Simulation and optimization of groundwater management strategies (operational boundary conditions, e.g., extraction regime, new usage).
3. Investigation of changing hydrogeological constraints (e.g., floods, droughts).
4. Worst case scenarios (geotechnical and hydrogeological).

Sometimes the complexity of environmental system is so immense that modeling approaches only result in general conclusions on regional groundwater flow regimes. This limitation is illustrated with a case study in Sect. 5.3 for an area of important drinking water production that is located in the vicinity of several contaminated sites. In such cases simple scenarios or scenarios that are described by conceptual models often are more feasible. However, for the derivation of regional groundwater budgets and flow velocities numerical modeling is indispensable.

4.3 Hydrogeophysics

Jannis Epting, Peter Huggenberger, and Edi Meier

Most urban aquifers are characterized by a high natural heterogeneity and a large spatial variability of hydraulic parameters (see also Sect. 4.4). Therefore, detailed knowledge of subsurface structures is an important prerequisite for the understanding and solution of specific problems related to adaptive resource management, in particular the qualitative aspects of groundwater and transport of contaminants. For a cost-effective exploration of the subsurface structures and hydrogeophysical properties, a combination of methods with different spatial resolution is necessary. The choice of the method depends on the specific goals of an investigation and the constraints given by the existing infrastructure.

In recent decades various hydrogeophysical techniques have been developed for mapping the subsurface. Hydrogeophysical methods thereby allow a spatial continuous characterization of the subsurface and can be used for a low-destructive mapping and monitoring. Hydrogeophysics has been successful in investigating and mapping complex geological areas in urban agglomerations also. Surveys can be performed in a relatively short time (days), taking into account different hydraulic and geotechnical boundary conditions, e.g., at low, average and high river discharge, before and after extensive construction measures. A number of geophysical techniques may potentially be applicable to investigations of geological structures near the surface, in boreholes (cross hole) or in combination of both. They can either portray the physical properties between the target and the surrounding media, different geological media or the change of physical properties of the fluids.

Each method has limitations in the depth of exploration and resolution, depending on their physical principles. The methods include, among others, noninvasive and minimal-invasive measurement methods, such as Electrical Resistivity Tomography (ERT), Ground Penetrating Radar (GPR) and ElectroMagentic Induction (EMI). Kirsch (2006) describes various hydrogeophysical applications for environmental sciences as well as hydrological, hydrogeological and geotechnical questions.

More often continuous hydrogeophysical measurements are conducted that also allow evaluating nonstationary processes such as, e.g., influencing hazards and groundwater pollution. For a fast characterization of heterogeneities of the subsurface the different hydrogeophysical methods can be combined with detailed vertical profiling. This can be accomplished by combining hydrogeophysical methods with direct-push technologies (Fig. 4.12). Table 4.1 gives an overview of hydrogeophysical methods that can be applied in urban areas, including spatiotemporal resolution and addressed investigation issues. Data deriving from hydrogeophysical investigation correspond to soft data (see also Sect. 4.1).

In this section we illustrate two hydrogeophysical sample applications. The first example demonstrates the application of underwater ERT conducted in the river bed upstream of a river dam. The results helped to delineate the thickness of sediment deposits behind the dam and to locate distinct karst features that promote preferential flow. The second example presents some results of a high-precision geodetic measurement system that was installed for monitoring displacement over large areas (Meier et al. 1998). We describe the application, development and refinement of the LAS (Large Area Settlement) hydrodynamic displacement monitoring system related to a slope stability problem on the river Rhine.

Fig. 4.12 Direct-push technology applied to investigate the subsurface and to install monitoring wells downstream of a landfill

Table 4.1 Hydrogeophysical techniques for urban areas and their field of application

	Field of application	Results
GPR	Detection of sedimentary structures, Depositional structures	Concepts for aquifer architecture and preferential flow (sharp boundaries)
Surface ERT	Detection of regions with similar electrical properties	Concepts for aquifer architecture and preferential flow (regions)
Underwater ERT	Detection of regions with similar electrical properties within the river bed	Concepts for riverbed architecture and infiltration windows (regions)
Crosshole ERT	Detection of regions with similar electrical properties between two boreholes	Concepts for aquifer architecture, fault structures, contaminant flow and preferential flow (regions)
EMI	Detection of regions with similar electrical properties	Concepts for shallow (<6 m) aquifer architecture, preferential flow and infiltration windows in rivers (regions)

4.3.1 Process Understanding

For process understanding the following specific research questions should be addressed:

1. How can multiple hydrogeophysical surveys, hydrogeological measurements, borehole information and profiles be combined to gain a comprehensive description of the subsurface structure and properties?
2. How can groundwater fluxes be estimated from discrete measured states?

Hydrogeophysical methods, e.g., infiltration monitoring, can be applied to determine the relevant processes of groundwater recharge, including river–groundwater interaction and near-surface processes. Such methods also incorporate:

1. Adaptive site characterization strategies, including the determination of site-specific relationships between hydrogeophysical and hydrogeological parameters.

2. The further development and evaluation of long-term monitoring concepts for the observation of river–groundwater interaction and near-surface processes as well as the investigation of the spatiotemporal resolution.
3. The application of hydrogeophysical techniques for long-term monitoring of water resources that allow real-time estimations of previously identified hazards (see Sect. 3.1).

In order to assess spatial infiltration patterns and their consequences on regional water balances nondestructive hydrogeophysical methods within the river bed and the riparian zone can be applied. They allow identifying zones of significant river–groundwater interaction within the gravel deposits. Such investigations can include surface and underwater ERT, GPR and EMI measurements.

4.3.1.1 Sample Application A

With the following example we illustrate some results from underwater ERT. The settings of the project are introduced in Sects. 5.2 and 5.4. For the location of the various profiles please refer to Fig. 5.11. The occurrence of the following two ERT anomalies is anticipated and used in these interpretations:

1. Karst features such as cavities, conduits, fractures and fault zones generally result in a electrical resistivity increase if these are filled with air (near-infinite electrical resistance), and a decrease if they are filled with clay and water, provided there is a electrical resistivity contrast with the surrounding rock. Although clay fractions will decrease electrical resistivity more than water, in the field, their influence cannot be determined due to the shape of the features and the fact that the degree of the filling is, most of the time, unknown.
2. Electrical resistivity contrasts between various sedimentological sequences and their degree of weathering.

Figure 4.13 shows the longitudinal profile combined with transverse profiles and the lithostratigraphic information of four boreholes. The results of the ERT measurements allow describing four distinct phenomena:

1. Zones with relatively high electrical resistivity values above 100 Ω m can be associated with sediment deposits behind the river dam. The rather high electrical resistivity values could be explained by the existence of coarse fluvial gravel. The thickness of the fluvial sediments range from 2 to 3 m and correlate with the lithostratigraphic information derived from the boreholes. The thickness of the sediment deposits increases from upstream regions towards the dam. These findings correlate with the progression of the surface of the gypsum rock (cf. Fig. 5.12).
2. The bottom of the sediment deposits corresponds to an erosion surface of the river preceding the dam construction. The undulating interface of low and high electrical resistivity values can be explained by the former progression of

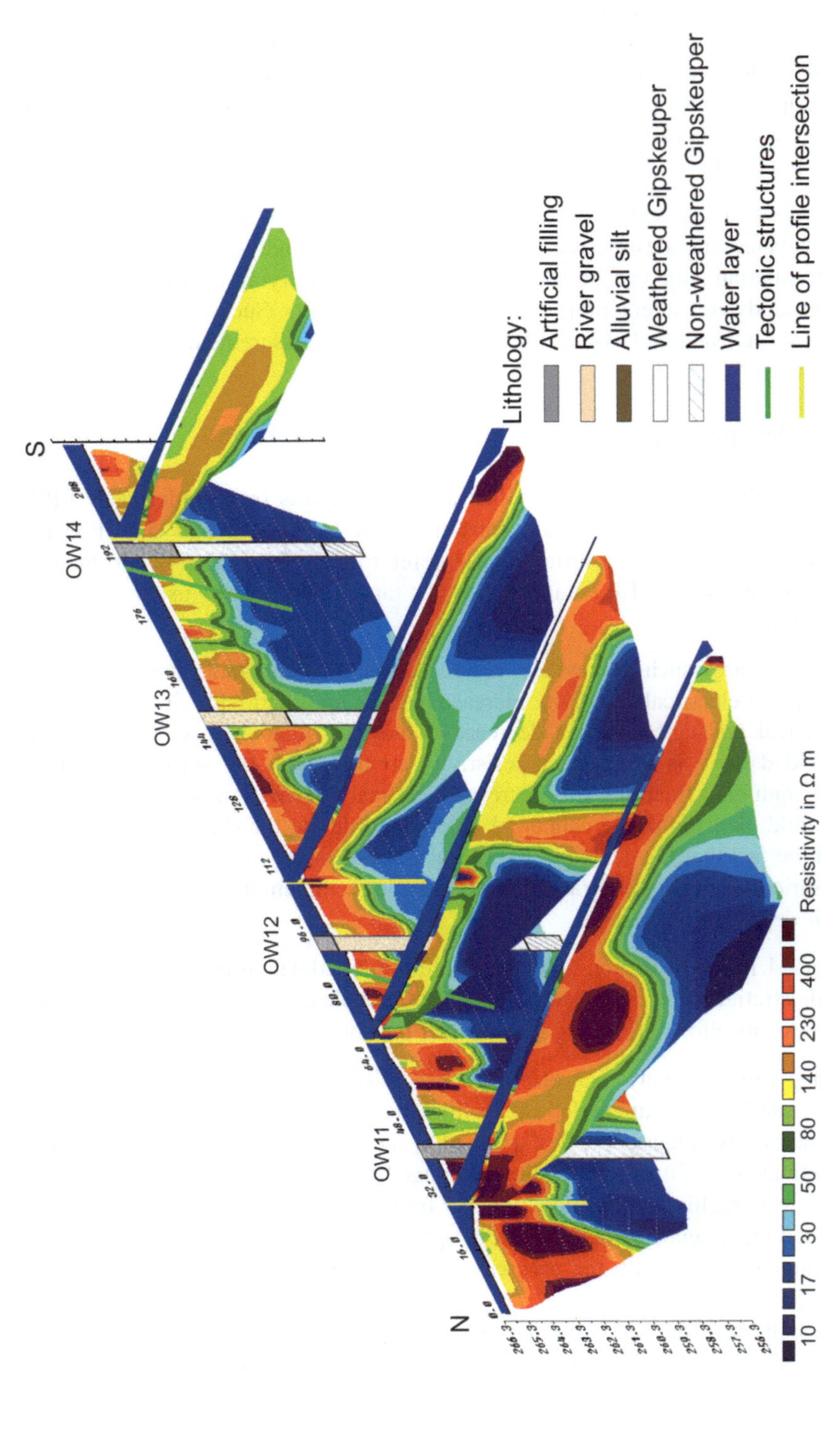

Fig. 4.13 3D view of underwater ERT measurements in the riverbed behind a dam illustrated together with lithostratigraphic information from boreholes

paleochannels and pool sequences. Findings accord with the results of the previous ERT measurements and information from the geological as well as historical maps.

3. Zones with relatively low electrical resistivity values up to maximal 40 Ω m beneath the sediment deposits can be associated to the weathered gypsum. Low electrical resistivity values result from water with high solution contents within water-saturated clays. Particularly within these zones, the weathering process resulted in the removal of gypsum and the remains of the clay component. Zones with higher electrical resistivity values between 40 and 100 Ω m on the western part of the first transversal profile can be explained by the more resistant Schilfsandstein. Findings correlate to the lithostratigraphic information derived from the boreholes at the river board and information from the geological map. As the underwater ERT measurements only reach to a depth of maximal 10 m, the detection of sharp boundaries between weathered and nonweathered zones was not possible.
4. Within the longitudinal profile at approx. 50 m, as well as in the second transversal profile at approx. 18 m, regions can be observed where the interpreted weathered gypsum rock is vertically cut through and high electrical resistivity values occur. An explanation for these structures beneath the river bed could be gravel-filled sinkholes, possibly also in combination with the local fault system. Similar karst features were observed in the riverbed further south. Furthermore, this karst feature can be part of a conduit system that reveals a siphon mechanism according to the hydrological characteristics of the river stage as has been described in Epting et al. (2009).

4.3.1.2 Sample Application B

In recent years various new geodetic measurement systems, such as GPS, motorized tacheometers and digital levels, have been introduced in permanent monitoring tasks. Most of these systems do not meet the submillimeter height accuracy which is achievable with hydrostatic or hydrodynamic systems. In addition, geodetic measurements, normally based on infrared and microwave methods, are affected by varying conditions of the atmosphere and troposphere and optical systems require a line of sight connection which can be obstructed by fog and other impacts during the operation.

On 4 February 1997, a slope failure destroyed a 20-m long section of a retaining wall of the quay on the left bank of the river Rhine at Basel (Fig. 4.14). The risk of eventual further movements causing damage to several historical buildings led to the decision to install a permanent survey system, to be able to recognize vertical components of land movements. A 22-point LAS meter was installed to monitor the zone continuously, provide information about the slide mechanism and also to predict the risk of a new slide. The LAS meter is based on a development which has been operating for several years to monitor a dam foundation (Meier 1991).

Fig. 4.14 The broken quay wall, a short time after the slide occurred

After the landslip the immediate priority concerned safety measures. This phase included the construction of a stable toe of the bank, with erosion protection, requiring the placement of some 900 tons of granite blocks. Then, the question arose as to whether the slip zone had stabilized or can movement continue. During this phase, experts put forward a number of possible causes for the incident and the destruction of the retaining wall. Some of the suggestions included:

1. Influence of the local geology, i.e., the complex topography of the bedrock below the unconsolidated rocks and the steep slope of the artificial fill.
2. Weakness of the retaining wall, which has been built as a pure gravity wall.
3. Effect of hydraulic collapse and inner erosion associated with small fracture systems in a waste-water pipe during an extraordinary high discharge 1996.

An accurate assessment of the cause was a primary requirement in the process of selecting a technical solution for the reconstruction of the retaining wall. In addition hydraulic and geotechnical studies aimed at gaining a better understanding of the processes involved, it was also important to conduct a survey of possible future movements. A combination of continuous LAS and periodical inclinometer measurements was considered to meet these requirements.

In March 1997, shortly after the slip occurred, a LAS meter was installed to monitor further ground motions. The LAS meter consists of one fixed central measuring unit attached to 22 measuring points. Perpendicular to the main planes of movement, two independent measurement chains were installed. Each chain consists of five measuring points (Fig. 4.15). The distance between adjacent points is about 2.5 m. The difference in height between the two points is a measure for the vertical component of ground movement. These results have been confirmed independently by a series of inclinometer measurements.

Conventional hydrostatic levels use the physical principle of communicating tubes and have several disadvantages, such as the need of connectivity, slow reaction or oscillation. The LAS meter (known as the hydrodynamic system) uses the pressure difference between two liquid columns. In the middle of the tubes a diaphragm is inserted which deforms according to the equalization of the liquid. As a result of pressure difference between the liquid columns, the diaphragm is deflected. The movement of this diaphragm is transformed to an electrical signal,

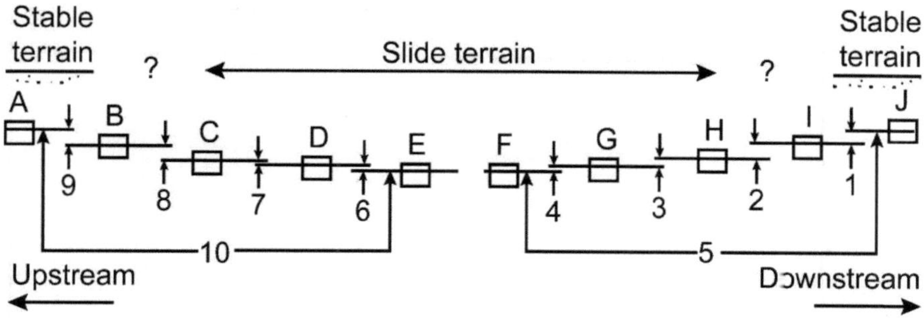

Fig. 4.15 Measuring concept for the control of the slide area. The measuring points E and F are in the middle of the slide areas and the points A and J are in the stable terrain on each side of the slide area; the numbers 1–10 represent the signals of the elevation differences recorded with the LAS meter (after Meier et al. 1998)

which is a measure of the level difference between the "chambers" at the end of the tube (Fig. 4.16). After installation, no further modifications to the instrument are necessary, because all operations can be carried out directly by the control unit. Full access to the measured data is possible via modem and telephone line or radio link.

Shortly after the instrumentation, the access to the buried measuring points was no longer possible because of the dense vegetation and the fact that the sliding area is private property. From each measuring point, three tubes lead to the central LAS meter, one for the measuring signal, one for filling the chamber and one for balancing local air-pressure differences. The complete system was filled from the central LAS meter. As shown in Fig. 4.15, the layout of the measuring system consists of two independent measuring chains. Chamber A is the reference point for the upstream chain, and chamber J for the downstream chain. The level difference between neighboring vessels is measured and sampled every 30 min. Each chain is closed with an overall measuring line for redundancy.

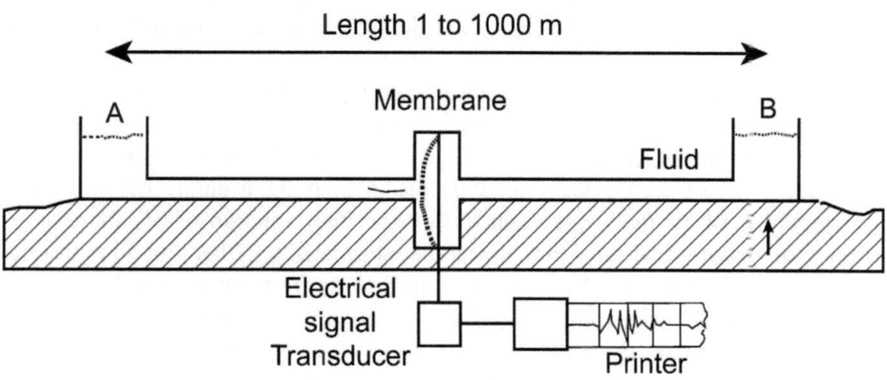

Fig. 4.16 Measuring principle of the LAS meter. As a result of the pressure differences between two fluid columns, the diaphragm becomes arched. The deformation of the diaphragm is transformed into an electric signal proportional to the elevation difference (after Meier et al. 1998)

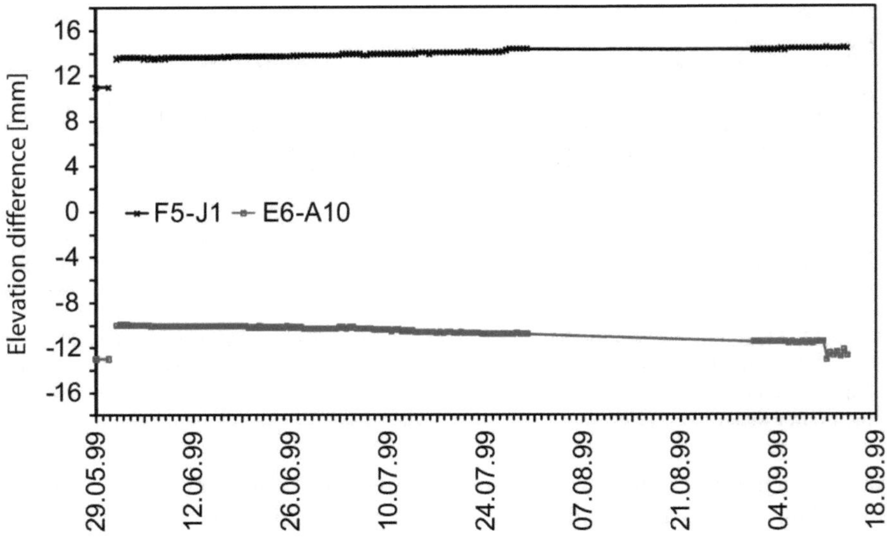

Fig. 4.17 Results of the redundant signals No. 10 and No. 5 displaying the elevation differences of the upstream chain A–E and the downstream chain F–J. Points E and F are in the middle of the slide area, A and J are set on stable terrain

Figure 4.17 gives an example of the quality of the data. The data set was recorded in September 1998. These data show the redundant measuring lines of the upstream chain A to E and the downstream chain F to J. The total tube length is 50 m for the upstream measuring line and 25 m for the downstream measuring line. The maximum deviation between the redundant lines is about 0.5 mm in the upstream and about 0.4 mm in the downstream measuring chain. Continuous recording at Basel has shown that the movement has stopped. However, the sensitivity of the system could be observed when remediation measures on the quay wall were started on 11 September 1999 and deviations were measurable within the upstream chain.

The non uniqueness of geophysical and hydrological parameter relations, scaling issues related to discrepancy of measured scale and process scale, and the site-specific suitability of hydrogeophysical methods are well known. Therefore the use of different hydrogeophysical methods results in more accurate definition and interpretation of anomalies. To consolidate the interpretation of results deriving from hydrogeophysical surveys, data can be interpreted together with (1) lithostratigraphic profiles from borehole logs; (2) information from outcrops and geological mapping; (3) monitoring data; as well as (4) hydraulics and water budgets derived from high-resolution 3D geological and hydrogeological models.

Hydrogeophysical surveying devices mostly do not directly measure the quantities of interest and there are still open questions related to scaling of the subsurface properties to field scale estimations of aquifer geometries. Thus the focus has to be placed on the understanding of the processes that are relevant and to develop concepts that include hydrogeophysical techniques to map and monitor proxies for the processes relevant to fluxes of water at the regional scale.

The implementation of hydrogeophysical techniques can provide basic datasets for process-based research regarding, e.g., river–groundwater interaction and fluxes of water. Since processes take place at different time scales, it is necessary to include short- and long-term monitoring strategies in the development of new technologies. This also includes the development and evaluation of combined investigation and interpretation approaches as well as the further development of hydrogeophysical methods for monitoring purposes.

Although today hydrogeophysical techniques are not used routinely, they will have a big potential in the near future. Results of hydrogeophysical investigations allow deriving 2D and 3D information for monitoring changes during engineering measures and to obtain site-specific information, e.g., on infiltration process. It is advisable that investigation focuses on locations where changes to urban subsurface resources occur. Thereby, the transition of the different investigation methods applies during planning and realization.

4.4 Aquifer Heterogeneity

Peter Huggenberger, Christian Regli, and Jannis Epting

Numerous urban areas in Central Europe and North America are located in flood plains. The alluvial valley fills are important aquifers for water resources on which water supplies strongly depend on. The complex fluvial and glaciofluvial depositional and erosional processes in these systems lead to highly heterogeneous distributions of hydrogeological parameters. This heterogeneity originates from sediment sorting processes in a dynamic depositional environment of aggradational and erosional processes typical of braided river deposits (Huggenberger and Regli 2006). Especially in urban areas, river systems have been channelized in the last two centuries.

Groundwater flow and solute transport processes within these coarse, permeable sediments are strongly influenced by subsurface heterogeneities and require detailed knowledge of aquifer properties, such as hydraulic conductivity, porosity and dispersivity, together with their spatial distribution.

On one hand, for most sedimentologists, the structure of aquifers can be assessed by understanding the depositional environments and by deriving general facies models that illustrate some of the physical conditions during deposition (e.g., Siegenthaler and Huggenberger 1993; Anderson et al. 1999). Hydrogeologists and engineers, on the other hand, are aware of subsurface heterogeneities; however, in their calculations, they often ignore or oversimplify subsurface heterogeneity, because of the difficulty to describe the complexity of natural systems. Although there was an enormous development in stochastic modeling techniques (e.g., Deutsch and Journel 1998; Renard et al. 2005), there is still a gap between stochastic approaches and conceptual sedimentological models that are appropriate to answer specific hydrogeological questions.

Often, the knowledge of specific flow paths within urban aquifer is not of major concern. This is the case for the drainage system of many construction sites, where engineers are mainly interested in quantitative information on how to lower the groundwater table during subsurface constructions (see Sect. 5.2). In addition, water suppliers generally are interested in quantitative aspects of water resources. However, experiences with ancient contaminated sites or emerging chemicals have shown that water quality issues suddenly may play a major role (see Sect. 5.3). In such cases, flow paths in the aquifers are essential for the evaluation of groundwater quality.

Many large engineering projects in urban areas require a temporal and location-specific lowering of the groundwater table, sometimes to the base of the unconsolidated rock. The often massive groundwater extractions produce extensive depression cones and associated capture zones to the drainage wells. As a result, these zones may interfere with old contaminated sites (see Sect. 5.2). The geometry and extent of the capture zones are related to the heterogeneity of the aquifer (Epting et al. 2008). Aquifer heterogeneity also plays a major role when considering the processes of river–groundwater interaction to design river engineering measures and to define groundwater protection zones (see Sect. 5.1, Regli et al. 2003). There are also efforts in semianalytical approaches which are based on estimations of uncertainty in the location of catchments of pumping wells due to the uncertainty of the spatially variable unconditioned hydraulic conductivity field (e.g., Stauffer et al. 2002).

In the present section we give a brief, nonexhaustive overview of some of the existing concepts to describe the character of aquifer heterogeneity in urban areas. We emphasize on the application of a sedimentological concept based on borehole descriptions, which we illustrate by a case study from the Basel area. This example not only illustrates a possible application but also addresses the problem of using borehole description as a database for aquifer structure studies.

4.4.1 Sedimentological Concept for the Description of Aquifer Heterogeneity

Hydraulic conductivity variations over several magnitudes are of key importance for groundwater flow and solute migration. Since continuous 3D information on hydraulic properties cannot be obtained in fluvial sediments, different methods have been developed to map aquifer properties. Koltermann and Gorelick (1996) distinguish three main types of methods:

1. *Structure-imitating methods* using any combination of Gaussian and non-Gaussian statistically and geometrically based relationships to match observed sedimentary patterns.

2. *Process-imitating methods* consisting of aquifer calibration techniques, which solve governing equations of fluid flow and transport, as well as geological process models combining mass and momentum conservation principles with sediment transport equations.
3. *Descriptive methods* using different field methods to translate the resulting geological sedimentary structure models into hydrofacies models with characteristic aquifer properties.

All these methods have already been applied in coarse glaciofluvial gravel deposits typical of braided river environments. However, for reasons of complexity of the aquifer structures in these coarse-grained sediments and effects of hydraulic property distribution, stochastic modeling was rarely applied to practical problems. The consideration of subsurface heterogeneity is often based on pumping tests, leading to a characteristically large-scale zoning of aquifer parameters. The generally scarce information on outcrop and the existing buildings and infrastructures preventing high-resolution hydrogeophysical investigations, as GPR, are important reasons for not considering heterogeneity in practical applications, particularly in urban areas. Therefore, borehole descriptions and partly sieve analyses with the known disadvantages concerning data quality and sedimentary structure recognition (Regli et al. 2002) are the most frequently available sedimentological data. Rarely indirect or soft data such as multilevel slug test data or flowmeter logs are also available.

A method of combined sedimentary structure and geostatistical analyses of borehole data was presented by Regli et al. (2002). The results of such analyses can be used to develop various simulations of stochastically generated aquifer properties, which can subsequently be integrated into multilayer high-resolution groundwater flow and transport models.

Information on the architecture of the aquifer is required to adequately model subsurface heterogeneity (Fig. 4.18). Outcrop and borehole information contains data of different quality and resolution at different scales. Outcrops of natural deposits above the groundwater table reveal distinct and coherent structural elements such as lenses and layers of different gravel types. Definition of sedimentary texture types is based on grain-size distribution and sediment sorting. Distinguished types of Rhine gravel are nicknamed according to color differences, including “brown gravel”, “grey gravel” and “gravel couplets”. Sedimentary structure types are made up of one or a combination of two, possibly alternating, sediment texture types. Sedimentary structure types include (Table 4.2): open-framework gravel (OW), open-framework/bimodal gravel couplets (OW/BM), gray gravel (GG), brown gravel (BG), alternating gray and brown gravel layers (GG/BG), horizontally layered or inclined, silty gravel (SG), sand lenses (SA), and silt lenses (SI). Important sedimentary structure types, such as the highly permeable OW, are generally overlooked due to smearing with overlying and underlying layers during the drilling process. Occurrence and size of OW determine, however, variance and correlation length of the hydraulic conductivity in coarse gravel deposits. Consequently, an important gap exists between outcrop and borehole descriptions. The strong association of OW to

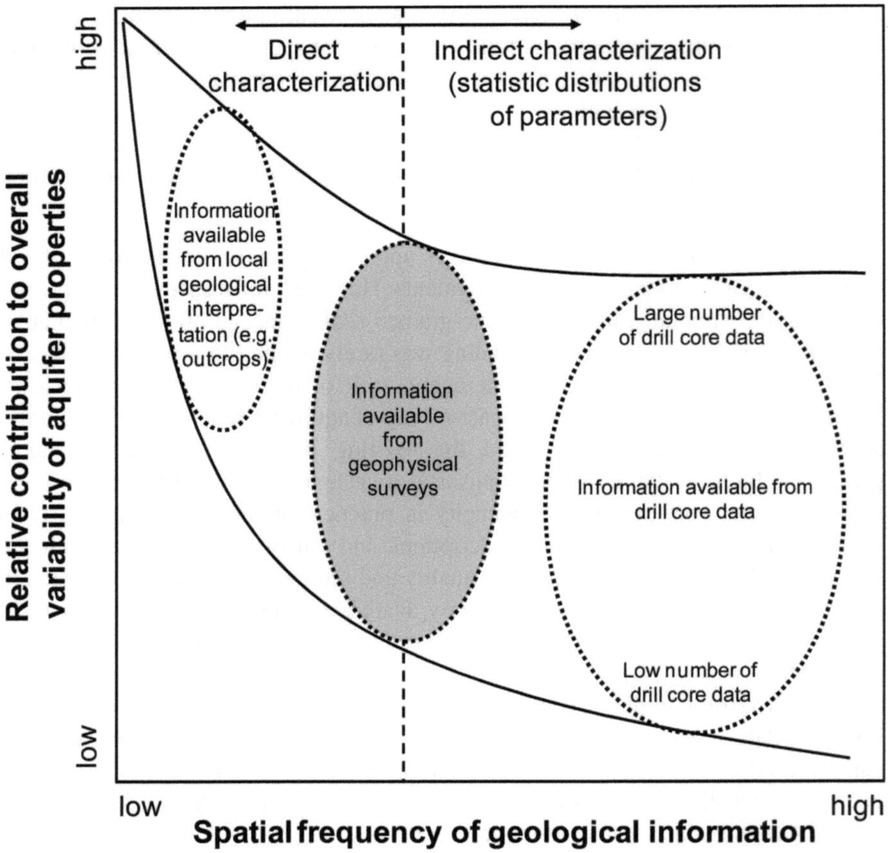

Fig. 4.18 The role of different kind of information for the characterization of aquifer properties

Table 4.2 Hydraulic parameters of sedimentary structure types used in the characterization of aquifer simulations (after Jussel et al. 1994)

	Sedimentary structure type								
	OW	OW/ BM	GG	BG	GG/BG horizontal	GG/BG inclined	SG	SA	SI
Hydraulic conductivity K (mm s^{-1})	100	10	0.15	0.02	0.08	0.1	0.008	0.26	0.005
Standard deviation $\sigma_{\ln K,1}$ (−)	0.8	0.8	0.5	0.6	0.8	0.8	0.5	0.4	0.4
Porosity n (%)	34.9	30	20.1	14.1	17	17	25	42.6	40
Local longitudinal dispersivity (mm)	25	30	25	30	30	30	3	0.3	0.05

the related structure type OW/BM has led to the concept of a gradual sedimentary structure-based interpretation of outcrop, borehole and GPR data. The method presented by Regli et al. (2002) allows a probability assessment of drill-core layer descriptions representing defined sedimentary structure types.

Fig. 4.19 Excavation pit enabling the definition of sedimentary structure patterns

Based on borehole information, defined sedimentary structure pattern for the investigation site (Fig. 4.19) and on the interpretation method for borehole data (see also Sect. 4.1), GeoData allows analysis of sedimentary structure and provides datasets of point information with arbitrary separation distances along boreholes. For groundwater modeling, the simulated sedimentary structure needs to be transformed into hydraulic parameters. The generated sedimentary structures are characterized by average and randomly selected hydraulic conductivity and porosity values provided by average and standard deviations calculated by Jussel et al. (1994). Files containing distributions of hydraulic conductivity, effective porosity and dispersivity values then are generated and exported to groundwater models. According to sedimentological and geostatistical analyses of the aquifer, each aquifer simulation corresponds to various equiprobable representations of the subsurface at variable degrees of uncertainty in hydraulic parameter values and geometry of the sedimentary structures.

Figure 4.20 shows the distribution of sedimentary structure types of a groundwater model covering an area of 450 m × 300 m (0.135 km^2; cf. Fig. 5.8). The spatial discretization resulted in total 1,350 cells of 10 m × 10 m cell size. For an appropriate vertical integration of aquifer heterogeneity, an approach with 13 horizontal layers was chosen. Each layer is 2 m thick and the total maximum vertical thickness amounts to 26 m. The interpolated surface of the aquifer base was then cut with the model grid, thus resulting in partly inactive cells in the two lower model layers. Finally, the interpolated aquifer properties (hydraulic conductivity, effective porosity and longitudinal dispersivity) were assigned to the prepared grid. A large-scale groundwater model (cf. Sect. 5.2) provided the boundary conditions. Model boundary conditions are of the first type (fixed head) along the northern side, and of the second type (fixed flow) along the eastern, southern and western side to account for the variable inflow and outflow rates across the boundaries. The eastern side covers residual leakage of the Rhine through the sheet pile wall, the southern side the groundwater flows beyond the tunnel construction and the western side focuses on the regional groundwater flow through this area.

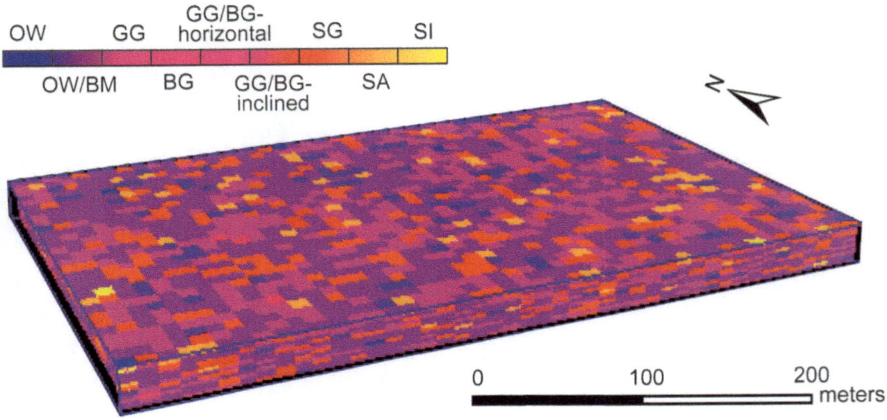

Fig. 4.20 Distribution of sedimentary structure types

Figure 4.21 illustrates the head distribution and flow paths for one set of modeled boundary conditions on 28 October 2004, assuming uniform distribution of aquifer parameters (left) and for a simulation with heterogeneously distributed aquifer parameters (right). The situation reflects the groundwater flow regime during construction of an emergency exit to a tunnel highway. Note the inflow running towards the construction site drainage. Integration of aquifer heterogeneities leads to an undulating progression of hydraulic heads.

Flow paths, visualized by particle tracks, reveal that high conductivity zones have a similar effect as optical lenses. The high hydraulic conductivity units (i.e., OW) are identifiable as they "focus" on the flow lines. Due to the complex interspacing of sedimentary structures, correlations between zones of high particle concentration and their associated sedimentary structures are visible only in a few places as illustrated in Fig. 4.21, i.e., in the center of the model domain where particle tracks are bundled when entering a high conductivity zone.

Note that although the hydraulic conductivity field is complex and heterogeneous, the resulting hydraulic head field is relatively smooth. In contrast, the local velocity field, as reflected by the movement of particles through the system, is quite complex and reflects more clearly the heterogeneity of the system.

To adequately evaluate potential mobilizations of contaminants, focus should be placed on aquifer heterogeneity. The applied techniques allow integrating data of different quality into groundwater models. Furthermore, with regard to contaminant transport on a local scale, the applied techniques present an approach to quantify the effect of groundwater flow budgets and velocities in the individual hydrofacies. Obviously, groundwater flow in heterogeneous media occurs largely through interconnected highly permeable geological aquifer structures. Together with hydrological and operational boundary conditions they govern the groundwater flow and transport regime. However, the relative amounts of groundwater budgets through the individual hydrofacies do not appear to significantly alter for

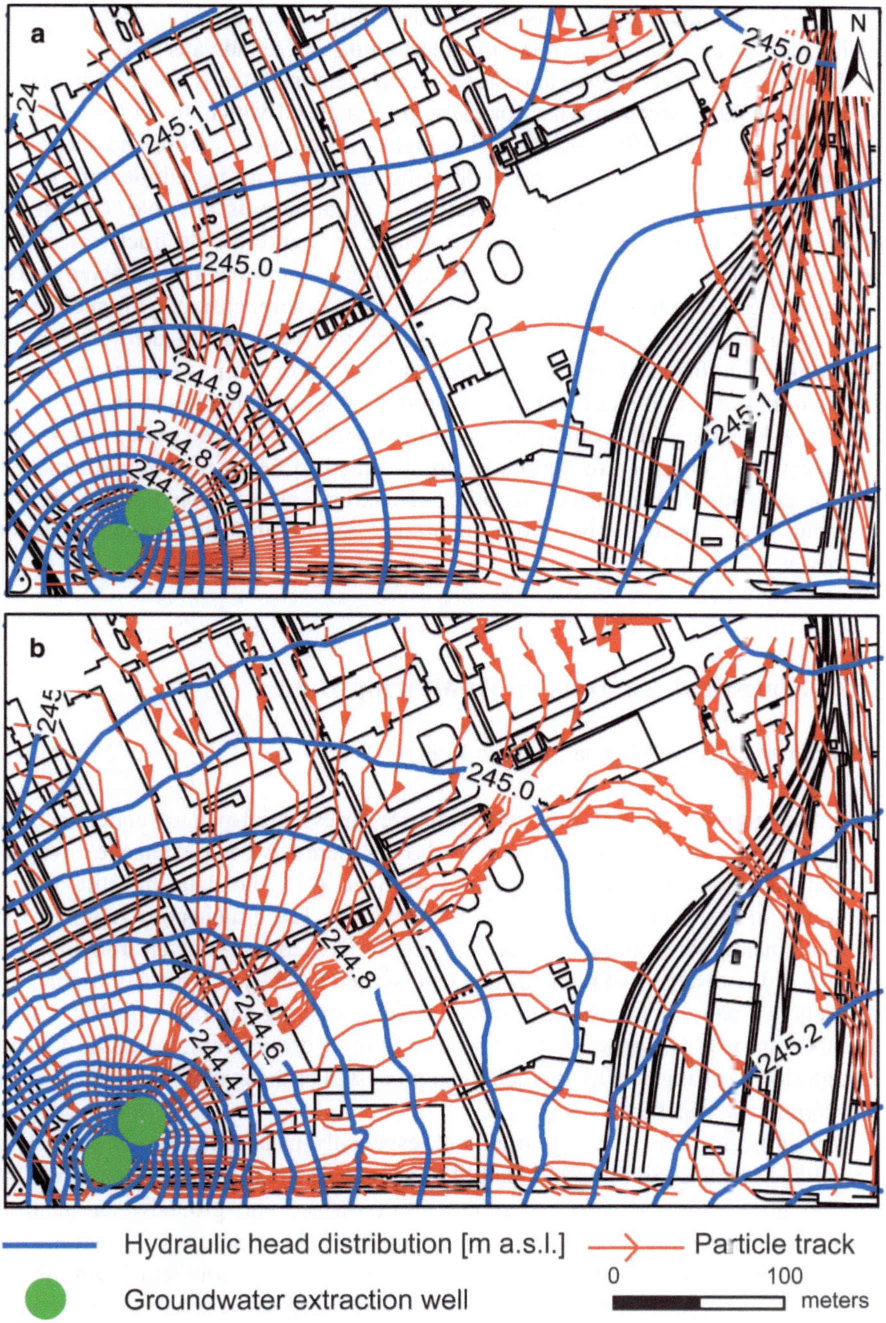

Fig. 4.21 Modeled groundwater flow regime without (**a**) and with (**b**) heterogeneously distributed aquifer properties. Visualization of flow paths illustrated by particle tracks (distance between two arrow heads indicates 100-day travel time). Topographic cards are reproduced with permission of swisstopo (BA110446)

the various boundary conditions investigated. Moreover, single hydrofacies and their relative occurrence determine the distribution of groundwater budgets.

The outcrop in the investigation area clearly demonstrated the size of the relevant sedimentary structure types for the region of Basel which are in the order of several tens to 100 m (Fig. 4.19). A lithological description (nature and shapes of sedimentary structure) is required to describe heterogeneity in sedimentary structure models and their resulting properties as accurately as possible. Optimized acquisition of geological recording of borehole data and less destructive drilling methods (drill cores in plastic liners) can significantly improve the characterization of sedimentary structure types. This includes more comprehensive hydrogeological investigations, i.e., a systematic collection and interpretation of drill cores as a function of lithofacies as well as hydraulic and hydrogeochemical parameters. Such innovations can involve tailored exports from geological databases such as separation of sedimentary lithocomponents into light and dark-colored components (see also Sect. 4.1). Since color variations are assumed to be an indicator of organic carbon content, they influence sorption capacities and sorption kinetics of the material. This information is of prior importance considering groundwater transport processes.

4.5 Statistical Analysis of Monitoring Data

Rebecca M. Page and Jelena Simovic Rota

Parameter monitoring and data collection are an essential part of understanding processes and ecosystem functioning. Although there has been extensive environmental data collection in urban areas (see Sect. 4.1), the data are still under-exploited in the management of natural resources. The conclusions required for decision making or defining management actions have to be extracted from the data collected. We discuss two methods that can be used to analyze large environmental datasets to obtain the necessary information for managing natural resources. These methods are readily applied to time series and offer considerable potential for the recognition of changes in hydrogeological processes or the evaluation of hazards in environmental systems.

A significant part of our environment, especially in urban areas, is subject to compliance with legal limits and quality standards. The compliance is measured by a substantial set of networks to monitor air, surface and groundwater quality. Mostly, guidelines are set for ideal and critical threshold values. The measurements are required to remain within the boundaries set by the threshold legislation. If the thresholds are breached, management actions are designed to return the system to the ideal state, e.g., groundwater extraction is ceased to protect the well. However, it may often be too late, as demonstrated by the collapse of the Northwest Atlantic cod fishery. As the cod stocks were severely depleted, the ban placed on the fishery was not sufficient for the stocks to recover. The demise in the system was recognized too late for the management actions to prevent a regime shift (Savenkoff et al. 2007). Besides the selection of an appropriate management action, the time at which it is employed is crucial.

Although not completely deterministic, environmental time series are not independent of predecessor or successor measurements. The development of a measured parameter over time is also strongly dependent on the dynamics of the system. Small, yet steady, changes can easily go undetected. They can, however, tell us a lot about system dynamics and indicate future development. The time scales at which these system dynamics-driven changes occur depend on the system under observation and the processes giving rise to the change. We have recently been confronted with the following specific questions: (1) how can we quantify the influence of climate change on the groundwater temperature in an urban environment (see Sect. 3.4); and (2) how can we use physico-chemical time series to assist in the assessment of actual hazard of microbial pollution for wells located near rivers. In both cases, large datasets make a simple analysis difficult. The information content of the dataset can, however, provide important guidelines for decision makers using data visualization methods (Fuertes et al. 2010).

Although the questions differ in their spatial and temporal scales, both require information on the development of the measured parameter over time in connection with observation points. In both cases, the preliminary analysis includes the assessment of variance in the dataset. In this context, the sources of variance are sought and the reaction of different observation points compared. Further analysis determines the development over time, the pattern arising from the variance. This aids the identification of situations where, e.g., river–groundwater interaction poses a threat to drinking water extraction, i.e., where a system moves away from a normal to a hazardous state. In other words, this section deals with pattern recognition in urban hydrogeological datasets. Applications of this methodology can also be found in Sect. 3.4.

These two case studies are the starting point for this brief overview of the application of principal component analysis (PCA) and a combination of self-organizing maps (SOM) with Sammon's Projection (SOM-SM) in urban hydrogeology. Detailed descriptions of the methodologies can be found in the literature cited.

4.5.1 Principal Component Analysis

PCA, also called the Karhunen–Loève transform, is a nonparametric method of extracting information from complex datasets (Lischeid 2009). It can be used to reduce the dimension of a dataset to reveal the sometimes hidden, simplified patterns.

The method is a linear transform based on correlation coefficients of the data matrix. The principal components (PCs) are uncorrelated (are orthogonal to each other) and represent the joint variance observed in the dataset (Gerbrands 1981). A PCA aims at representing a maximum fraction of variance by a small number of components (Lischeid 2009). In simple systems one to two components may be sufficient to summarize the major sources of variance. The more complex a system is, the more components are required to explain the observed patterns in the data. There are different criteria for the optimal number of components to interpret (e.g., Kaiser 1960; Cliff 1988). The loading of an input vector (original data) with a component provides an estimation of the importance of the source of variation (given by the component) at each observation point or for each measurement (input vector). A PCA can thus be carried out on continuous time series or discrete samples. Helena et al. (2000) studied the influence of temporal variations, e.g., caused by precipitation or agricultural activity, on groundwater quality by applying a PCA to a large number of discrete samples. Lewandowski et al. (2009), on the other hand, carried out a PCA on continuous groundwater head measurements to study the relationship between groundwater head and river stage over time. The example provided in Fig. 4.22 shows the analysis of groundwater temperatures in an area influenced by river–groundwater interaction and intensive groundwater use. The variance explained by the first two PCs is compared to potential sources of variation, such as air temperature and river stage fluctuations. The correlation between the input dataset and the PCs provides the analyst with an impression of the influences on the temperature recorded at each observation well with respect to the sources of variation identified (Fig. 4.22a). The degree of influence of the two major factors on groundwater temperature in this example is given by the location of individual observation wells around the unit circle (Fig. 4.22a). The observation wells can be characterized by their sources of influence in terms of susceptibility to surface water fluctuations or longer term, seasonal variation.

Many datasets in the field of urban hydrogeology are heterogeneous. They often derive from specific monitoring networks associated with construction or maintenance of urban infrastructure and are therefore temporally or spatially patchy and closely related to the problem at hand, e.g., temperature plumes around buildings or electrical conductivity measurements close to construction sites. While a PCA can provide a means of comparing multiple parameters simultaneously, it is applied to a finite dataset. Each analysis is dependent on the input dataset and small changes will potentially result in a different outcome, it can be considered a stationary method. The development of system state with time cannot be sufficiently represented to be used in a decision-making process, which requires a time-sensitive approach.

4.5.2 *Artificial Neural Networks*

The automatic detection of events, such as hazardous states requires a time-sensitive approach. PCA will provide an indication of the distribution of influence of processes, but not how the system as a whole develops over time. Artificial

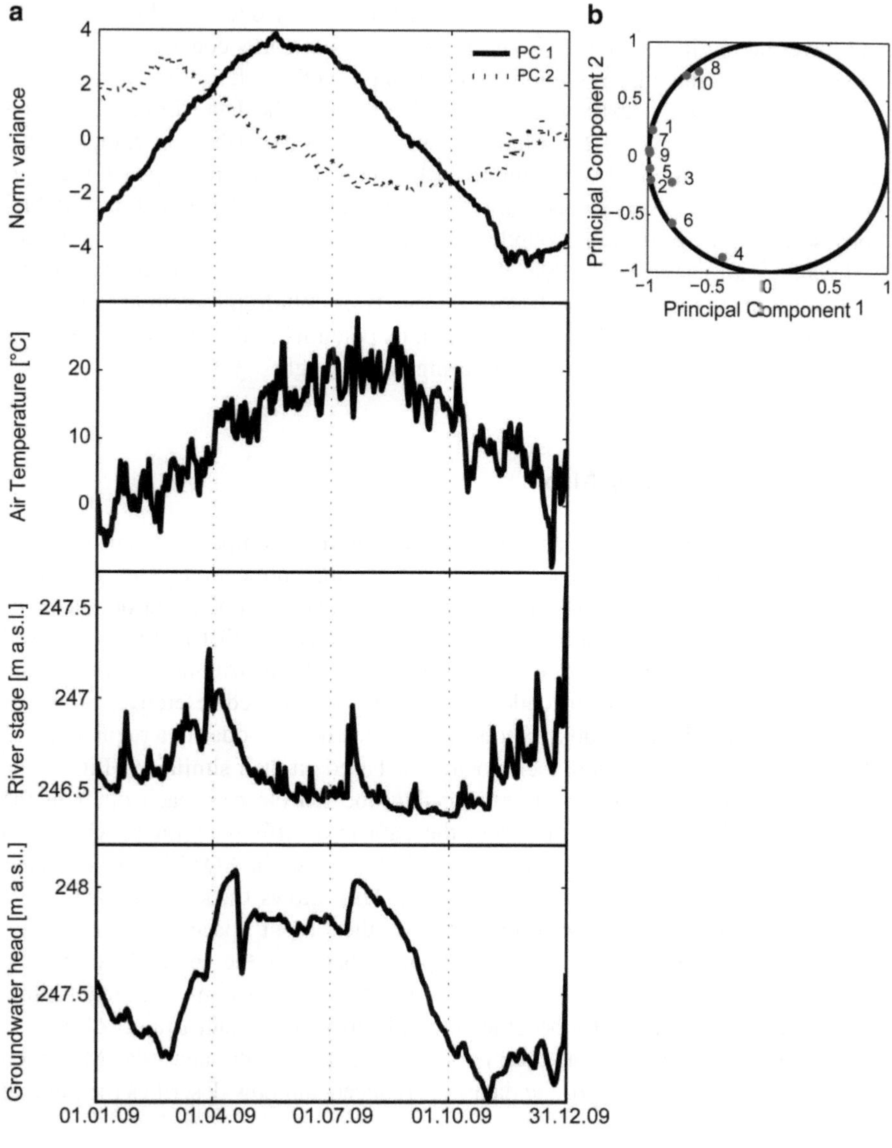

Fig. 4.22 (**a**) Time series of PC1 and PC2 shown in connection with two major sources of influence on temperature variation during 2009: air temperature and river stage fuctuations. The *bottom graph* shows the head time series of a groundwater observation well (number 9, location see Fig. 3.20) mostly influenced by PC1. (**b**) Unit circle representation of the influence of PC1 and PC2 on the time series of each observation well. The closer a point is to the rim, the more variance is explained by PC1 and PC2. In comparison to observation well 9, observation well 4 is more dependent on PC2 suggesting river–groundwater exchange is a driver for variation at observation well 4

neural networks (ANN) are increasingly used to classify data based on similarity and have the ability to "learn" from the data. The networks consist of a series of nodes that are defined from the input dataset and functions describing the relationships. ANNs are able to extract knowledge directly from the data without an explicit physical model by resolving nonlinear input–output relationships in complex systems. ANNs have a broad spectrum of applications, ranging from speech recognition over image analysis to anomaly detection. ANNs can be used for modeling purposes, e.g., the impact of groundwater extraction on rivers or predicting microbial water quality (Lin et al. 2008). However, the focus of this overview lies on the application of the methods for time-series analysis. While the method is increasingly applied to survey industrial processes, its potential in the domain of resource management and environmental monitoring is very high.

4.5.2.1 Self-Organizing Map

The SOM (Kohonen 2001) is an ANN method based on competitive, unsupervised learning. SOMs have found use in many engineering applications, e.g., for monitoring industrial process states or drinking water quality in distribution networks (Dominguez et al. 2007; Mustonen et al. 2008; Corona et al. 2010). SOMs are based on vector quantization where an approximation to the distribution of input data vector is made using a set of codebook vectors (also called reference, model or weight vectors). The codebook vectors are associated with nodes in a regular grid, a 2D or 3D output space, where their position is based on their similarity (Fig. 4.23). Similar situations, or process states, are thus located close to each other in the output space. This leads to clustering the data into different groups, which can then be classified, e.g., as "normal" or "hazardous." As the SOM is a visual data-mining approach (Dominguez et al. 2007), it also allows tracking of the process state by visualizing trajectories projected to the output space. This occurs by selecting the node with the least discrepancy between the input data and the codebook vector, the best-matching unit (BMU), for each measurement time. The succession of the BMUs becomes the trajectory. The result is a set of vectors describing the system state at each point in time used in the analysis (Mustonen et al. 2008). The dimension of the dataset is reduced and now describes the internal structure of the data matrix.

4.5.2.2 Sammon's Mapping

Sammon's mapping is a nonlinear mapping algorithm. It is aimed at preserving the distances in the measurement vector in a 2D projection (Sammon 1969). Sammon's mapping is very useful in determining the shape and density of clusters and the relative differences between these clusters (Kolehmainen et al. 2003), which cannot be represented by a SOM. In Sammon's mapping, the cells are located so that the

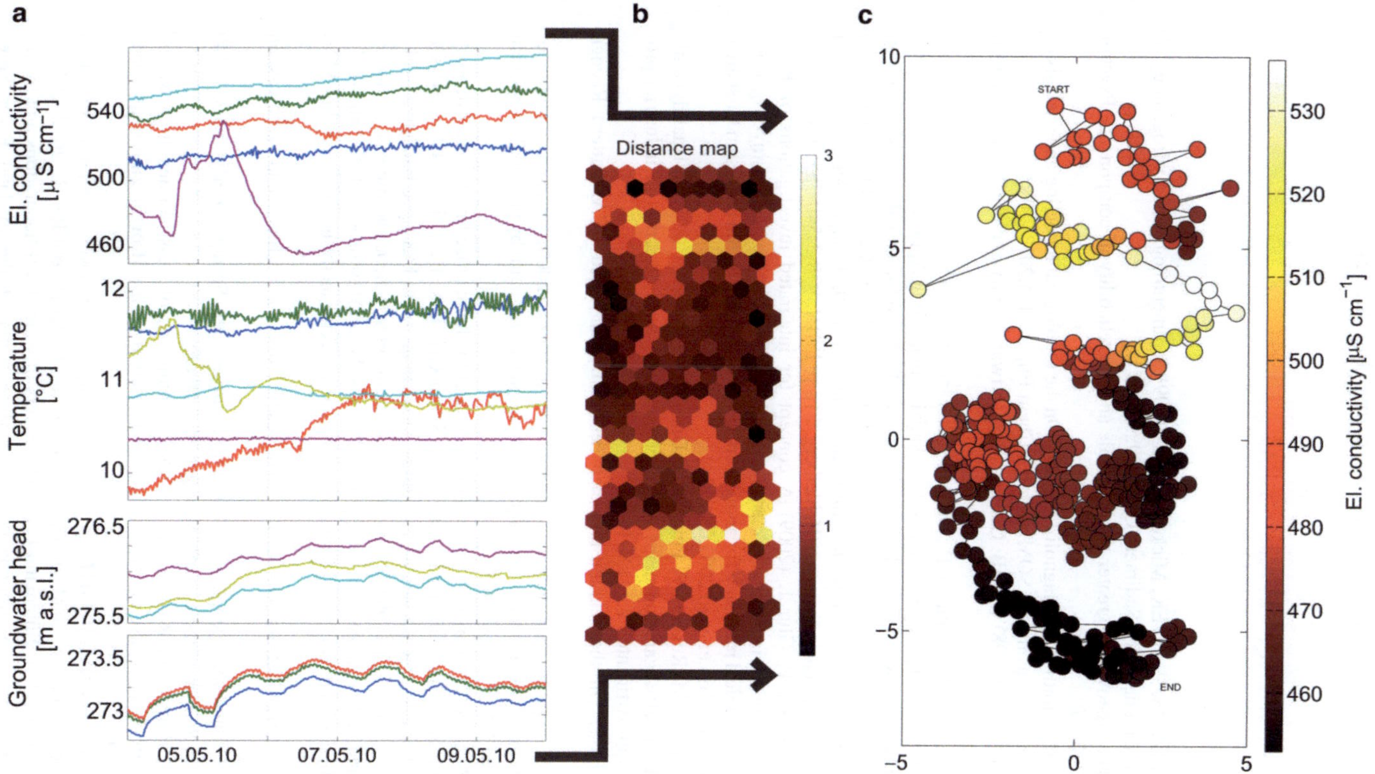

Fig. 4.23 Steps in data mining with multivariate groundwater time series of a dataset with 17 variables. The wealth of data is shown in panel (**a**), where time series in electrical conductivity, temperature and groundwater head recorded in up to six observation wells show different amounts of variation. Panel (**b**) shows the output from the self-organizing map analysis: a distance map. The distance refers to the similarity, or dissimilarity (*shading light*: dissimilar, *shading dark*: similar), between situations. Panel (**c**) includes time in the projection, so each point visualized represents an individual measurement situation. The development of the dataset (all 17 variables) can be tracked through time. The *shading* of the points is given by electrical conductivity measurements from one observation well

distance between them represents dissimilarity, providing information about the change of a system under consideration over time (Fig. 4.23c).

References

Anderson MP, Aiken JS, Webb EK, Mickelson DM (1999) Sedimentology and hydrogeology of two braided stream deposits. Sediment Geol 129:501–511

Cliff N (1988) The eigenvalues-greater-than-one rule and the reliability of components. Psychol Bull 103(2):276–279

Corona F, Mulas M, Baratti R, Romagnoli JA (2010) On the topological modeling and analysis of industrial process data using the SOM. Comput Chem Eng 34(12):2022–2032

Deutsch CV, Journel AG (1998) GSLIB: Geostatistical software library and user's guide. Oxford University Press, Oxford

Dominguez M, Fuertes JJ, Reguera P, Diaz I, Cuadrado AA (2007) Internet-based remote supervision of industrial processes using self-organizing maps. Eng Appl Artif Intell 20(6):757

Epting J, Huggenberger P, Regli C, Spoljaric N, Kirchhofer R (2008) Integrated methods for urban groundwater management considering subsurface heterogeneity. In: Cai X, Jim Yeh TC (eds) Quantitative information fusion for hydrological sciences, vol 79. Springer Series: Studies in Computational Intelligence. Springer, Heidelberg, 218 p. ISBN: 978-3-540-75383-4

Epting J, Huggenberger P, Glur L (2009) A concept for integrated investigations of Karst phenomena in urban environments – merging geophysical and hydrometrical investigations with 3D hydrogeological modeling for applied urban hydrogeology within a gypsum karst area. Eng Geol. doi:10.1016/j.enggeo.2009.08.013

Fuertes JJ, Dominguez M, Reguera P, Prada MA, Diaz I, Cuadrado AA (2010) Visual dynamic model based on self-organizing maps for supervision and fault detection in industrial processes. Eng Appl Artif Intell 23(1):8–17

Gerbrands JJ (1981) On the relationships between SVD, KLT and PCA. Pattern Recogn 14:375–381

Helena B, Pardo R, Vega M, Barrado E, Fernandez JM, Fernandez L (2000) Temporal evolution of groundwater composition in an alluvial aquifer (Pisuerga River, Spain) by principal component analysis. Water Res 34(3):807–816

Huggenberger P, Regli C (2006) A sedimentological model to characterise braided river deposits for hydrogeological applications. In: Sambrook-Smith GH, Best JL, Bristow CS, Petts GE (eds). Braided rivers: process, deposits, ecology and management. IAS Spec Publ 36:51–74

Jussel P, Stauffer F, Dracos T (1994) Transport modeling in heterogeneous aquifers: 1. statistical description and numerical generation of gravel deposits. Water Resour Res 30(6):1803–1817

Kaiser HF (1960) The application of electronic computers to factor analysis. Educ Psychol Meas 20:141–151

Kirchhofer RT (2006) GeoData. Geological Data-base of the city of Basel and northwestern Switzerland

Kirsch R (2006) Groundwater geophysics. A tool for hydrogeology. Springer, Berlin, pp 321–340

Kohonen T (2001) Self-organizing maps. Springer, Heidelberg

Kolehmainen M, Ronkko P, Raatikainen A (2003) Monitoring of yeast fermentation by ion mobility spectrometry measurement and data visualisation with self-organizing maps. Anal Chim Acta 484(1):93–100

Koltermann CE, Gorelick SM (1996) Heterogeneity in sedimentary deposits: a review of structure-imitating, process-imitating, and descriptive approaches. Water Resour Res 32:2617–2658

Lewandowski J, Lischeid G, Nützmann G (2009) Drivers of water level fluctuations and hydrological exchange between groundwater and surface water at the lowland River Spree (Germany): field study and statistical analyses. Hydrol Process 23:2117–2128

Lin B, Syed M, Falconer RA (2008) Predicting faecal indicator levels in estuarine receiving waters – an integrated hydrodynamic and ANN modelling approach. Environ Model Softw 23(6):729–740

Lischeid G (2009) Non-linear visualization and analysis of large water quality data sets: a model-free basis for efficient monitoring and risk assessment. Stoch Environ Res Risk Assess 23:977–990

Meier E (1991) A differential pressure tiltmeter for large-scale ground monitoring. Water Power Dam Construct 43:38–40

Meier E, Huggenberger P, Ingensand H (1998) Precision monitoring of displacement over large areas. Hydropower Dams Issue 6:77–80

Mustonen SM, Tissari S, Huikko L, Kolehmainen M, Lehtola MJ, Hirvonen A (2008) Evaluating online data of water quality changes in a pilot drinking water distribution system with multivariate data exploration methods. Water Res 42:2421–2430

Noack T (1993) Geologische Datenbank der Region Basel. Eclogae Geol Helv 86:283–301

Pahl-Wostl C (2006) Newsletter No. 1. http://www.newater.info

Regli C, Huggenberger P, Rauber M (2002) Interpretation of drill-core and georadar data of coarse gravel deposits. J Hydrol 255:234–252

Regli C, Rauber M, Huggenberger P (2003) Analyses of aquifer heterogeneity within a well capture zone, comparison of model data with field experiments: a case study from the river Wiese, Switzerland. Aquatic Sci 65:111–128

Renard P, Demougeot-Renard H, Froidevaux R (2005) Geostatistics for environmental applications. Springer, Heidelberg, 480 p

Sammon JW (1969) A nonlinear mapping for data structure analysis, IEEE Trans Comput C 18(5):401–409

Savenkoff C, Castonguay M, Chabot D, Hammill MO, Bourdages H, Morissette L (2007) Changes in the northern Gulf of St. Lawrence ecosystem estimated by inverse modelling: evidence of a fishery-induced regime shift. Est Coast Shelf Sci 73:711–724

Siegenthaler C, Huggenberger P (1993) Pleistocene Rhine gravel: deposits of a braided river system with dominant pool preservation. In: Best JL, Bristow CS (eds) Braided rivers. Geol Soc Spec Publ 75:147–162

Stauffer F, Attinger S, Zimmermann S, Kinzelbach W (2002) Uncertainty estimation of well catchments in heterogeneous aquifers. Water Resour Res 38(11):1238. doi:10.1029/2001WR000819

Chapter 5
Examples and Case Studies

Peter Huggenberger, Jannis Epting, Annette Affolter, Christoph Butscher, Donat Fäh, Daniel Gechter, Markus Konz, Rebecca M. Page, Christian Regli, Douchko Romanov, Stefan Scheidler, Eric Zechner, and Ali Zidane

The examples and case studies we illustrate from the region of Basel, Northwestern Switzerland, focus on questions with practical as well as research character in the domain of urban geology. Although the proposed content covers particularly regional projects, most topics are relevant for urban areas and the sustainable use of subsurface resources in general.

As our key objective we implement the sustainability concepts which we discussed in Sect. 3.1. Such concepts together with the setup of tools and process-oriented experiments allow testing hypotheses. The overall results of the applied methods and concepts allowed us to fill several gaps about our knowledge of subsurface processes. The various examples and case studies integrate existing and new scientific knowledge into new methods, concepts, and tools that allowed us to better manage the subsurface resources in the Basel area.

In a first set of case studies we address protection issues of groundwater production and river restoration in urban areas, with a focus on drinking water supply aspects. We present protection schemes for several major drinking water supplies in the region of Basel. We focus on hydrogeoecological issues in the context of river restoration projects in urban environments. Urbanization in the last century created a series of environmental problems such as flooding, groundwater pollution, and ecological changes, including a decrease of characteristic habitats of riverine landscapes together with a drastic reduction of species. With three examples we illustrate strategies to integrate hydrogeoecological aspects in an early planning process of engineering projects, which include drinking water and flood protection measures as well as river restoration in urban areas. Further we focus on the setup of monitoring networks and modeling tools, river–groundwater interaction, aquifer heterogeneity, and the reconciliation of water engineering measures along rivers.

In a second set of case studies, we address engineering and hydrogeological questions that emerged during the development of urban infrastructure projects in the region of Basel. Here we focus on groundwater management and protection issues during and after completion of two infrastructure development and upgrading projects.

P. Huggenberger and J. Epting (eds.), *Urban Geology: Process-Oriented Concepts for Adaptive and Integrated Resource Management*, DOI 10.1007/978-3-0348-0185-0_5,

In a third set of case studies, we encompass management concepts as well as monitoring, modeling, and remediation strategies for contaminated sites in transboundary settings. In a first case study, we discuss strategies to understand and predict the cumulative effects of the numerous single impacts on groundwater resources during a major subsurface development project. In a second case study, we illustrate the development of groundwater pollution in a heavily industrialized groundwater protection area during the last decades.

In the fourth set of case studies, we address karst in urban environments. Groundwater circulation in evaporite bearing horizons and the resulting evolution of karst frequently causes geotechnical problems such as land-subsidence or collapses. Such processes are of particular concern in urban areas where soluble geological formations coincide with vulnerable infrastructures as transportation systems. In this section, we focus on two case studies where subrosion, land-subsidence, and impacts on infrastructures have been observed. The case studies allow the illustration of the complex interrelations between natural phenomena and processes that are induced by present day engineering and subsurface activities in the region of Basel.

In the fifth set of case studies, we address the use of shallow geothermal energy in urban environments. Increasing geothermal energy use can exceed the subsurface potential for different heating and cooling systems and effect groundwater quality. Currently, in most urban areas, regulations for water resource management and geothermal energy use are sparse and limited to the rule "first come, first served." In this section, we focus on concepts for monitoring and modeling the influence of geothermal systems as well as on the provision of suitability maps for site evaluation. In the first case study, we present a concept that allows to rapidly evaluate proposed drilling sites that are suitable for geothermal use. In the second case, we present a thermal groundwater management concept on the basis of developed monitoring and modeling tools.

In a sixth set of case studies we deal with natural hazards in a regional context, including major flood events and flood protection measures as well as earthquakes and earthquake risk reduction.

5.1 Groundwater Protection and Hydrogeoecology

Peter Huggenberger, Jannis Epting, Christian Regli, Annette Affolter, Stefan Scheidler, Christoph Butscher, and Rebecca M. Page

In Switzerland, about 40% of drinking water is derived from gravel aquifers in river valleys, which are to a great extent urbanized. The river valleys in the alpine forelands and the Upper Rhine and Bresse Grabens are the most important groundwater reservoirs in Central Europe. These reservoirs are very vulnerable due to the intense agricultural activities and the dense network of urban areas which are connected by a large number of traffic lines. The lack of a thick protective cover layer, the high permeability of the fluvial sediments, and the exchange processes with surface waters make groundwater resources very vulnerable. The integral of changes in river structure in the catchment (lack of retention space) already had serious consequences during larger floods (River Rhone, Switzerland, or river Elbe, Germany).

With over 40% of the water supply of Western and Eastern Europe and the Mediterranean region coming from urban aquifers, efficient and cost-effective management tools for this resource are essential to maintain the quality of life and economic development, but also to ensure that water is available for future generations in sufficient quality and quantity (Eiswirth et al. 2003). Sustainable use of soil and groundwater resources and protection and conservation of their quality are hence a key issue of European environmental policy and an enormous challenge for European research (Prokop 2003). Urbanization comes along with increasing pressure on regional water resources (quality and quantity) and together with rising energy demands will lead to more use conflicts in the future.

Hydrogeoecology in urban areas may be considered as an interdisciplinary branch of hydrogeology that integrates the surface and subsurface hydrological, hydrogeological, as well as ecological aspects of riverine systems. A particularity of urban areas is the interference of processes in river valleys with the development of urban infrastructures. Currently, the focus of most infrastructure projects is on the engineering side and projects are not dimensioned according to long-term sustainable aspects of a particular riverine urban landscape.

In this section, we investigate urban groundwater bodies (GWB) and their usage related to groundwater protection and the interference during flood events as well as water engineering activities along rivers. Our concept for groundwater protection and adaptive groundwater management (see Sect. 3.1) is illustrated by three examples from two rivers in the agglomeration of the city of Basel (Fig. 4.10). We further focus on river restoration projects and processes of river–groundwater interaction.

5.1.1 Current Status of Urban River Valleys

Channelization of most rivers in Switzerland and in many parts of Europe in the nineteenth and twentieth century strongly influenced the economical development. Together with impacts of urbanization this resulted in a series of environmental problems such as flooding, groundwater pollution, and ecological changes. Characteristic habitats of riverine landscapes decreased and species were drastically reduced. Because of its values for recreation and leisure activities the restoration of rivers becomes an important element in the planning of cities agglomeration and infrastructure.

Due to the experience gained from hazardous flood events in the last 20 years, the pollution problems and the lost of characteristic riverine landscapes, most countries have acquired a more comprehensive view of rivers. It was recognized that hazardous floods cannot be controlled by increasing the height of dams because of hydraulic base failure.

It is also widely accepted now that measures for contemporary flood protection have to consider the catchment scale and the allocation of space for rivers required to mitigate the effects of hazardous flood events and to provide rivers the required dynamics.

Most rivers can be subdivided into segments according to the dominance of either in- or exfiltration and changing conductance properties of the riverbed. The interaction thereby is subject to permanent dynamics considering water budgets, water quality, and flow patterns. Riverine groundwater consequently has not a uniform and constant physical, chemical, and biological signature. The composition can temporally vary significantly depending on the dynamics of particular systems. In addition, groundwater quality may be degraded due to sporadic impact loads from surface waters, i.e., caused by urban storm water drainage or by effluents from sewage treatment plants.

Current research confirmed that groundwater recharge in many river valleys, especially in northwestern Switzerland, is dominated by infiltrating river water (see Sect. 3.2). In addition, in many Swiss river flood plains that are used for drinking water production, artificial recharge of river water is an important groundwater recharge component (see Sect. 3.2).

The dynamics of the different recharge processes are important factors for understanding aquifer habitats of a natural biocenosis that interact with surface waters. Rivers with natural dynamics exert a dominating influence on erosion, transport, and sedimentation processes. As a consequence, the variance of riverbed permeability can be increased temporarily, influencing infiltration rates, groundwater mixing ratios as well as residence times of groundwater of different provenance. Therefore, water engineering projects along rivers require careful and comprehensive evaluations considering riverine groundwater and its usage. A detailed, site-specific understanding, including the consideration of various hydrological boundary conditions as well as evaluations of riverine groundwater and its usage, is the basic requirement for water engineering measures along rivers.

Formulated goals for a sustainable development of water systems guide mitigation strategies and refer to defined standards, i.e., natural composition of surface waters and groundwater or quality standards defined by existing regulations. They also establish a standard that has to be defined at different scales, including the definition of critical control points (see Sect. 3.3), against which individual decisions are made. The base for defining goals for future transdisciplinary projects with respect to the hydrological regimes (low flow and high flow) or groundwater flow regimes is the knowledge of the physical properties governing the hydrogeo-ecological systems. General goals, among others, could be, e.g.:

1. Enhancement of the interaction between surface water and groundwater.
2. Maintenance of local and regional groundwater flow regimes.
3. Quality-oriented surface water and groundwater management.
4. Consideration of future water and groundwater use.
5. Long-term improvement of surface water and groundwater quality.

There are now techniques available to identify and quantify system profiles for hydrogeological cycles at different scales (see Sect. 3.1). This can be achieved by the setup of management systems that involve observation systems as well as the setup of numerical models combined with scenario development. Besides a simple documentation of changes in water quantity and quality, the goal of the management systems is to predict desired and undesired developments in advance. As a basis for rational decision making in urban water management, this includes the ability to make predictions on consequences of proposed measures such as the:

1. Influence of flood protection measures.
2. Changes in agricultural activities at local and at basin scale.
3. Restoration of natural dynamics of particular river sections.
4. Influence of global change, as e.g., increasing temperatures (see Sect. 3.4).

5.1.2 Main Changes from the Natural to the Channelized State of Rivers

Since the channelization of rivers two centuries ago several important factors influence riverine groundwater flow regimes. Some of these factors are related to the basic processes of river–groundwater interaction and influence exploitable aquifers in urban environments. They include:

1. *Formal separation of surface and subsurface water systems:* In the nineteenth and far into the twentieth century, the channelization of rivers progressively limited the transversal and vertical interconnectedness of rivers with their floodplains and groundwater together with a reduction of the thickness of the hyporheic zone. Furthermore, the longitudinal connectivity (river continuum) between the various river reaches and the main river was restricted. In many places this led to an entire loss of the natural dynamics of river systems. Channelization measures led to uniform flow patterns and increased peak floods. Bank and bed protection prevented the erosion of sediments and, therefore, relocation within the active channel belt. Mostly this has led to a lack of bed load and a clogging of the riverbed and the interstitial. As a result, river–groundwater interactions are reduced along with a decreased filtration of surface water in the pore space of gravel beds (Kozel 2005; Regli and Huggenberger 2006).
2. *Protection properties of the overlying stratum (protective soil cover):* In the context of river–groundwater interaction, the basic concept of protection capacity of the soil cover is not valid. In riverbeds the soil cover is missing and the infiltration rates can show strong spatial and temporal changes according to the thickness and conductance of the riverbed, the structure and permeability of the river bank, and the relationship between flow depth and groundwater table. In addition, in the presented examples river–groundwater interaction can be reduced or enhanced by riverine groundwater extraction or artificial recharge (Regli and Huggenberger 2007).

3. *Groundwater residence times:* In the past, groundwater extraction wells were often constructed very close to rivers. The reason for the site selection near river banks was high conductivities and storage properties that were favorable for drinking water production. Nowadays, the proximity to the rivers is often disadvantageous, because groundwater residence times and filtration capacities are often below or close to threshold values. During flood events, a part of the infiltrated river water stays only a few days in the subsurface before it enters the extraction well (Hoehn 2005).
4. *Groundwater mixing ratios:* Due to the different infiltration rates, the mixing ratios of riverine groundwater and groundwater with longer residence times are controlled by dynamic changes. The consideration of transient infiltration and the resulting changes in groundwater mixing ratios during different hydrological conditions enhance the understanding of the interaction processes between surface and subsurface water systems. Furthermore, they are a necessary basis for estimating the risk of pollution for riverine groundwater and its usage, resulting in site-specific, adequate protection measures and adapted groundwater management strategies for extraction wells.
5. *Filter capacity between river and extraction well:* The elimination of particles in the subsurface passage due to filtration, sorption, and biochemical processes is the determining factor for the microbial quality of groundwater. These processes mainly occur in the soil and subordinate in the nonsaturated and saturated zone (BUWAL 2004).

5.1.3 Reconciliation of Water Engineering Measures Along Rivers

An enhanced reconciliation of the various usage demands with groundwater protection issues includes, along with aspects concerning water quality and quantity, the restoration of rivers in their function as species-rich ecosystems that form landscapes and interlink different habitats. Water engineering measures along rivers have to be accomplished to mitigate the impact of hazardous flood events and the conservation and recovery of natural functions of water systems. Therefore, such projects have to incorporate the interests of qualitative, quantitative, and ecological groundwater protection issues. Our experience from different river restoration projects illustrates the need for the definition of goals for specific river restoration measures (i.e., sufficient space for rivers, sufficient discharge and reasonable water quality; BUWAL 2004) and a fine tuning with those of groundwater protection, considering the fact that water engineering measures along rivers might not only locally affect the groundwater system, but can also influence the groundwater flow regime, the groundwater quality, and the ecology downstream. The spatial context is hence not only restricted to the vicinity of planned impacts on water systems, but can often concern system dynamics covering large areas of the floodplain (Huggenberger et al. 2006).

In general, decisions to compensate for negative impacts are often made at the level of the individual project. The effect of mitigation policy on the groundwater flow regime in urban areas depends in part on how the regulatory community defines "equivalence" (see Sect. 3.1). The basic premise of compensatory mitigation is that measures taken compensate for, or at least reduce, the effects of local damage. However, cumulative effects of water engineering measures could have an influence at considerable distances from the specific impact location. This required the development of modeling tools that facilitate to adequately quantify the consequences of cumulative effects arising from the numerous decisions concerning the groundwater flow regime and groundwater quality.

The strategy of the development of modeling tools with predictive character in the domain of groundwater management systems is illustrated based on three examples from the Basel area and the INTERREG III Project Monit (see Sect. 3.2). These tools may be refined and calibrated for different scales and extended with socioeconomic tools allowing focusing atomized responsibilities.

5.1.4 Endangerment and Hazard Assessment

A schematic illustration of the quality of infiltrated river water (e.g., a substance concentration in the river) against the filter performance in the region between the riverbed/foreland and the extraction well allows the designation of different areas (Fig. 5.1; Regli and Huggenberger 2007). The illustration facilitates the formulation

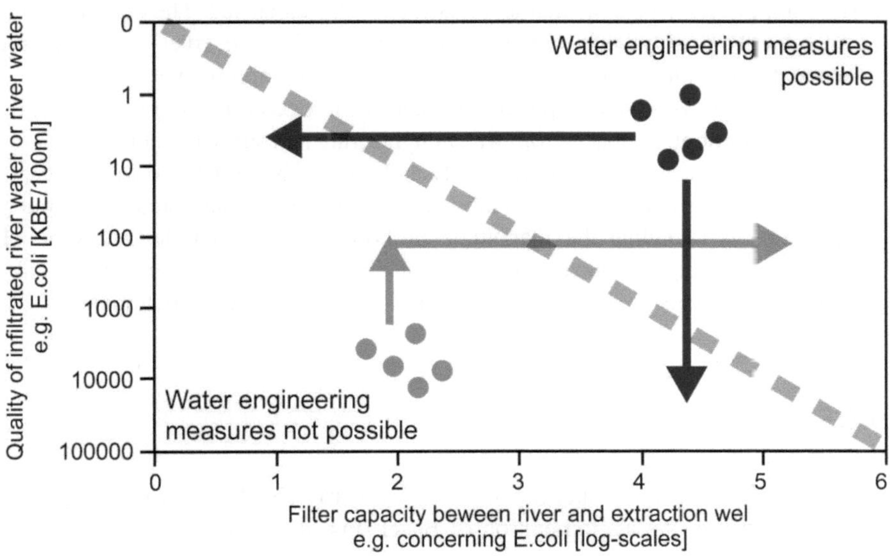

Fig. 5.1 Conceptual diagram for the evaluation of water engineering measures along rivers. *Dotted line*: definition of target size, e.g., adherence of threshold values in the extraction well

of requirements for water engineering measures along rivers. The separation of these areas is defined by a line, marking the threshold value of a substance (dotted line, exceeding a threshold value or substance concentration). In a planning process the choice of one or several parameters (e.g., *E. coli*), which are considered for safety evaluations of drinking water supplies, is determined by "problematic" substances (i.e., emerging chemicals, Wells et al. 2007) in the river or in the catchment areas. The filter performance, defined as the ratio of the substance concentration in the extracted groundwater to that in the infiltrated river water or river water, is particularly dependent on the load of the river water, the structural properties of the riverbed and the aquifer (infiltration rates, groundwater mixing ratios, residence times) and groundwater flow patterns, as well as the properties of substances and of the groundwater.

Due to strong heterogeneities of aquifers and the transient character of hydraulic conditions during flood events the attenuation of specific compounds or particles can vary considerably. Water qualities below the line that marks the threshold value for a specific substance in the drinking water require measures for quality improvement. When the quality of infiltrated river water or river water is degraded (impact loads during flood events and/or decreased filter capacity) the threshold line could also be undercut (dark gray horizontal and vertical arrow). If for a riverine extraction well, a scatter plot of water quality can be characterized that lies beneath the threshold line (quality objectives for drinking water cannot be achieved), the quality of the infiltrated river water or river water and/or the filter performance have to be improved (light gray arrows). This would result in a better chance and higher degree of freedom to facilitate water engineering measures. Furthermore, it should be considered that when improving the ecological state of rivers, the filter performance of the riverbed and the interstitial is also enhanced. The quantification of the achieved quality improvements, however, is difficult. In river segments with permanent exfiltration, water engineering measures are unproblematic.

The challenge is to define protection goals with a basic reflection on possible hazards. Accordingly, impacts from water engineering measures and flood events for riverine groundwater usage can effectively and efficiently be reduced to an acceptable degree. A way to assess the risk of water engineering measures and forthcoming flood events is to define the magnitude of floods that restrict the operation of extraction wells (e.g., frequency of events or discharge quantities, maximum substance concentrations in rivers, and duration of accepted usage restrictions).

5.1.5 Case Studies

A combination of groundwater modeling and scenario development is exemplified by case studies in the agglomeration of Basel (Fig. 4.10). To define the specific profiles of groundwater systems high-resolution groundwater models are applied that have been calibrated with time series of groundwater head data and river stages as well as groundwater extraction and recharge rates. In the presented examples, the strongly transient character of river–groundwater interactions in urban areas is illustrated. Scenario techniques have been developed to assess consequences of possible decisions and to optimize particular measures such as channel widening and their influence on groundwater quality.

Three examples illustrate the strategies to integrate hydrogeoecological aspects in an early planning process of infrastructure development in an urban area in the domains of drinking water and flood protection as well as river restoration. The examples focus on river–groundwater interaction, quality-oriented groundwater monitoring, as well as adaptive groundwater management, and consist of the following three case studies (1) groundwater modeling and scenario development along the river Wiese, suggesting differentiated solutions when considering river restoration in urban areas; (2) data analysis from extensive monitoring programs along the river Birs, including the development of transient riverbed conductance models and their implementation into high-resolution groundwater modeling systems (cf. Affolter et al. 2010); and (3) groundwater modeling and scenario development, suggesting river restoration as prerequisite for bringing the beaver back to the headwaters of the Birs River.

The first case study is from the floodplain of the river Wiese near the confluence of the river Rhine and covers an area of about 6 km^2. The second and third examples are from the river Birs in the Lower Birs valley, and cover an area of about 12 km^2, bounded by tectonically influenced higher ground to the East and to the West. In both areas the drinking water supply competes with other interests and demands such as river training, flood control, recreation, and change of land use.

5.1.5.1 River Training and Flood Protection

Many engineering projects along rivers that could affect riverine groundwater production lack efficient groundwater protection concepts. Also it has to be accepted that, in particular cases, changes in river structures cannot be completed without endangering groundwater quality. The multitude of strongly transient processes makes risk assessment for particular well locations difficult. A clear definition of the present groundwater system profile, including its transient character, would help to define realistic goals and targets for site-specific conditions (see Sect. 3.1). Considering the bandwidth of possible solutions, from the abandonment of riverine groundwater wells to the foregoing of corresponding interferences to water systems, there should be options of adequate measures. In some cases the goal is to work out options that provide adequate space for groundwater usage as well as for river systems. These challenges increase the requirements for investigation and assessment methods. The first example illustrates the application of scenario techniques for evaluating solutions for river restoration, with emphasis on conflicts with groundwater protection issues.

Before entering the river Rhine, the river Wiese flows through its former floodplain that widens up towards the river Rhine. Whereas the position of the active channel migrated considerably at earlier times, the river bank is fixed since

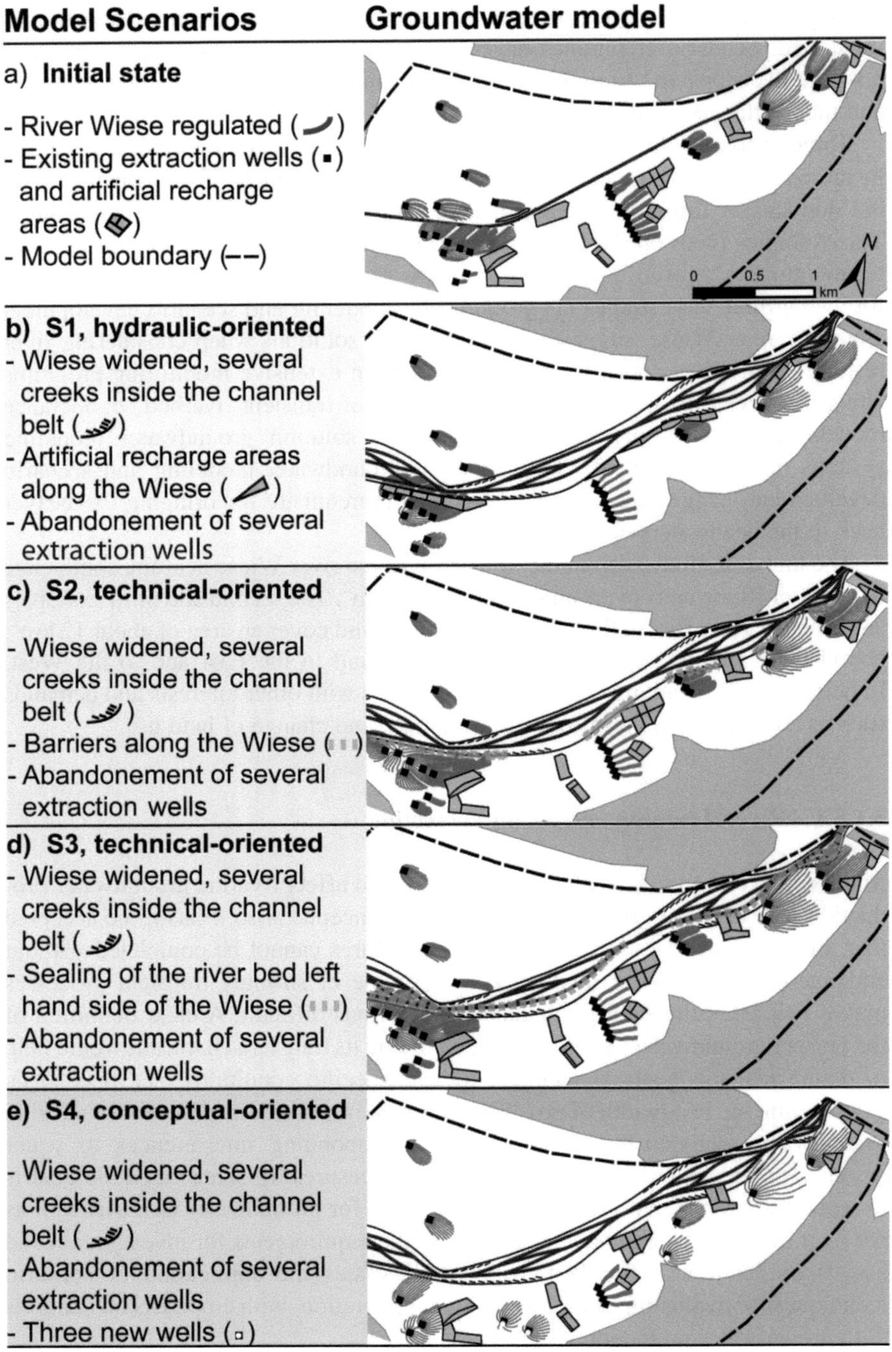

Fig. 5.2 Model scenarios and capture zones of groundwater extraction wells in the Wiese floodplain (10-day-period including a flood event)

150 years. Due to the vicinity to major urban areas (Basel, Switzerland as well as Lörrach and Weil, Germany) the flood plain area of about 7 km^2 is primary used as groundwater production area. The plans to reconnect the headwaters of the river Wiese with the river Rhine and to provide habitat for salmon was the main reason for an ongoing controversy on river restoration versus groundwater protection. This was the starting point to setup a groundwater monitoring system together with a high-resolution groundwater model for the whole area. Based on these models we determined the present profile of the aquifer, including the present risk originating from surface–groundwater interactions (Fig. 5.2).

In a second step, the modeling of scenarios with different degrees of freedom with respect to river dynamics and river–groundwater interactions allowed to evaluate possible groundwater flow regimes. The scenarios were grouped into conceptual-, technical-, and hydraulic-oriented scenarios (Regli et al. 2004). Examples of conceptual-oriented scenarios are (1) investigation, planning, and dimensioning of groundwater extraction areas; (2) enhancement of surface water quality; (3) optimization of urban drainage, e.g., sewage drainage into larger receiving streams; (4) reduction of water consumption, which would allow some of the extraction wells to be abandoned; and (5) planning and investigating alternative well locations. Examples of hydraulic measures and their influence on the river–groundwater interaction are illustrated schematically in Fig. 5.3. By adequate arrangement and operation of groundwater recharge areas, hydraulic barriers can be generated. Thereby, groundwater recharge areas would function as temporarily wetted floodplain surfaces. However, minimum groundwater residence times and/or minimum infiltrating water quality should be ensured according to defined threshold values and to achieve drinking water standards in extraction wells. Examples of technical measures and their influence on the river–groundwater interaction are illustrated schematically in Fig. 5.3. The insertion of sealing walls in the vicinity of riverine groundwater wells results in vertical barriers that prevent the infiltration of river water into the aquifer. A more reasonable ecological alternative could be technical solutions such as geo-textiles that decrease infiltration rates as well as amounts of fines and reduce seepage velocities at certain segments of the river extent. Calculated scenarios include the consideration of several minor creeks with adapted infiltration capacities, the relocation of extraction wells and groundwater recharge areas (Fig. 5.3). For the evaluation of the various scenarios, the capture zones of groundwater extraction wells were evaluated and compared. For the consideration of conceptual-oriented scenarios, several wells located proximal to the river were abandoned and new wells at locations more distant to the river were introduced. This allowed the influence of inflow from infiltrated river water to be reduced.

Figure 5.2a shows the profile of the groundwater system in its initial state. Figure 5.2b illustrates the hydraulical-oriented scenario including an alternating operation of possible groundwater recharge areas along the river Wiese. Optimization of capture zones can be achieved by alternative arrangement and operation of artificial groundwater recharge areas and the location of extraction wells in combination with the consideration of the river hydrograph and flood events. Figure 5.2c, d shows the capture zones of the extraction wells, when considering technical-

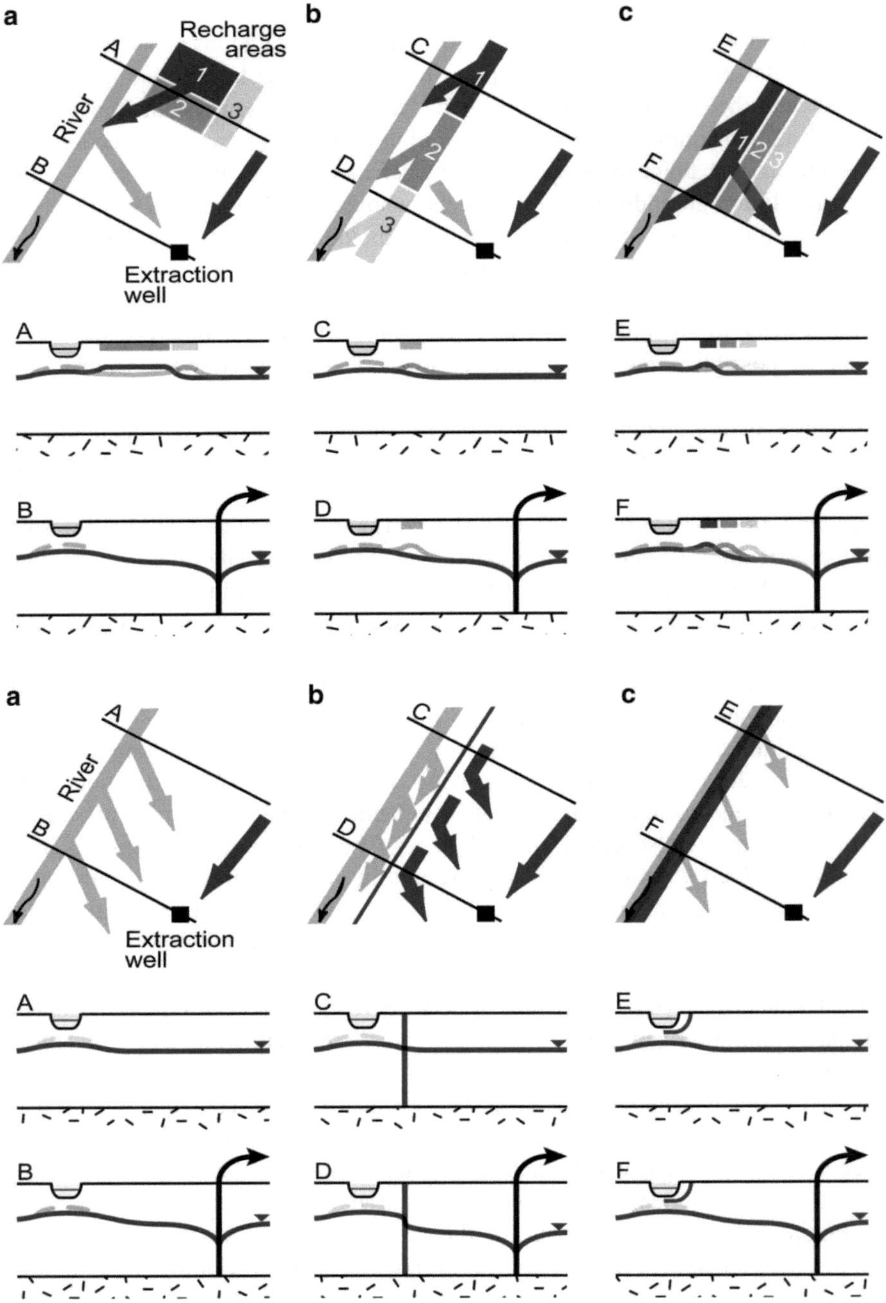

Fig. 5.3 *Above*: Hydraulic measures and their influence on river–groundwater interaction: (**a**) Current status in the Wiese floodplain; (**b**) groundwater recharge areas parallel to the river; (**c**) several groundwater recharge areas parallel to the river. *Numbers* on the recharge fields indicate alternating time periods of recharge operation. *Below*: Technical measures and their influence on river–groundwater interaction: (**a**) Current status in the Wiese floodplain; (**b**) vertical barriers parallel to the river; (**c**) horizontal barriers of single river segments

oriented scenarios with vertical and horizontal barriers. The conceptual-oriented scenarios shown in Fig. 5.2e illustrate the influence of reducing and relocating groundwater extraction wells.

In an early planning phase, there are several ways to increase the degree of freedom for rivers. Our experience is that the evaluation of different options is time consuming. Thereby, the discussion on the required drinking water is part of the planning process and might be addressed at a regional level including quality and quantity aspects to be coordinated between different supply systems. Measures on the catchment scale include, e.g., the optimization of settlement drainage and to target an improved river water quality. Considering organizational and technical arrangements on extraction wells, such as updated concessions, linked systems, adaptation of groundwater extractions in relation to discharge combined with load variations (impact loads) or the shutdown of extraction wells (adaptive groundwater management), UV-installations, etc., the smallest irreversible constructional measures are necessary. Possible technical measures are, e.g., the adaptation of the planned interferences in the river, the relocation of riverine extraction wells (enlargement of the filter passage and consequently the filter performance), the injection of groundwater or the installation of geo-textiles (hydraulic and technical barriers, changing the groundwater flow regime). All these measures require elaborate reconcilement among the authorities in charge of the environmental and technical aspects (Regli and Huggenberger 2007).

5.1.5.2 River Training and Flood Protection

The second example illustrates current profiles of a groundwater system in an urban environment that is influenced by artificial groundwater recharge and river–groundwater interactions as well as agricultural and industrial activities. Extensive analytical groundwater monitoring programs during and after a water engineering project allowed the definition of particular profiles of the groundwater system. This was accomplished by the setup of a transient groundwater model and the evaluation of various scenarios in the Birs valley (Münchenstein, Switzerland; Fig. 5.4). The technical measures focus on flood protection and the protection of the river bank (erosion) as well as on an ecological reassessment of the river Birs. Therefore, a 250 m section of the river board will be restored. Additionally, a groundwater extraction well is located within 50 m of the river. However, the proposed changes should not degrade the quality of extracted groundwater. The groundwater models allowed defining critical river reaches and capturing zones of wells during different hydraulic conditions. Based on this information construction measures are proposed that reduce, or at least do not increase, the infiltration of river water into groundwater. A monitoring concept is proposed that allows detection of changes to the composition of raw water quality in the extraction well during the construction phase. The concept comprises continuous groundwater monitoring in the extraction well by incorporating measuring sensors (electrical conductivity, turbidity/particles,

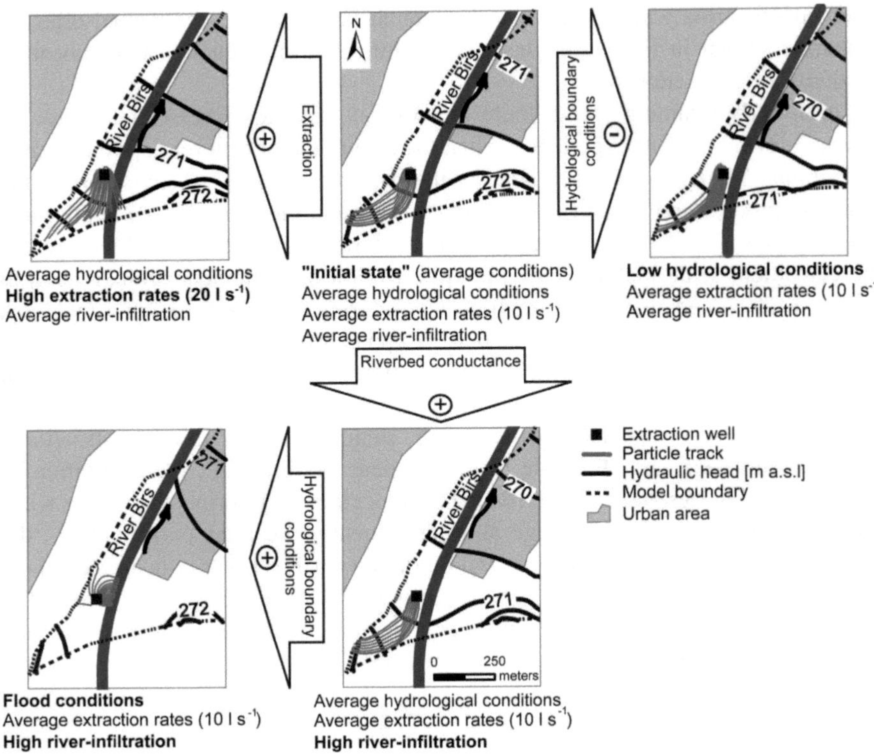

Fig. 5.4 Groundwater modeling and scenario development in the Birs valley (+ increase, – decrease)

UV-extinction, temperature) that should allow detection of the signature of infiltrated river water that only remains a few days in the subsurface. Furthermore, the monitoring program includes extensive analysis of the raw water for selected microbiological contaminations before, during, and after the water engineering measures.

For groundwater modeling and the developed scenarios, different hydrological and operational boundary conditions are considered. Based on average hydrological boundary conditions, average extraction rates (10 l s^{-1}), and average river infiltration rates, several boundary conditions were changed to evaluate the influences on the capture zone of the groundwater extraction well (Fig. 5.4):

1. For overall average boundary conditions, the inflow to the extraction well is mainly from the agricultural area to the southwest. This is supported by groundwater quality data (high nitrate and microbiological content).
2. When elevating the groundwater extraction rates, the capture zone is widened and includes parts of the river Birs. This might reduce the nitrate concentration but elevates the risk of microbiological impacts.

3. When considering low hydrological boundary conditions, less groundwater is derived from the agricultural area to the southwest while more comes from southern areas. This should not change groundwater quality significantly.
4. When elevating the riverbed conductance the capture zone of the groundwater extraction well moves away from the river. Considering average hydrological and operational boundary conditions as well as an elevated riverbed conductance, more groundwater exfiltrates into the river (and no river water infiltrates into the groundwater). Thus, no effect on the groundwater quality is expected.
5. During flood events the extracted groundwater is derived from short passages to the river. Thereby significant microbiological vitiations can be expected.

5.1.5.3 "Hello Beaver" (Bringing Back the Beaver to the Headwaters of the River Birs)

The third example illustrates a project that was initiated in the late nineties by nongovernmental organizations (NGOs) to bring the beaver back to the headwaters of the river Birs. Based on these initiatives, the adjacent communities propose a restoration of a 1 km river reach and to integrate the whole floodplain into an existing recreation area. The river restoration project originates from local and regional environmental actors as well as natural preservation. The goal to bring the beaver back to the headwaters is shared by most habitants in the area, including the regional and local governments. However, the consequences for planning and realization in an urban context lead to use conflicts with other competing leisure or sport projects and with respect to groundwater production.

The community, in charge of the planning and the environmental organization, came up with questions about suitability of different project options. Especially the change of river–groundwater interaction along three different sections of the 1 km river reach had to be evaluated, also with respect to the transient character of capture zones of two existing wells. We evaluated four different scenarios and the resulting groundwater flow regime (1) present situation; (2) minimum restoration; (3) medium restoration; and (4) maximum opening (Fig. 5.5). For these scenarios the infiltration rates from the river to the groundwater and capture zone of wells were calculated and compared (Fig. 5.6).

The results of the model calculations show that the capture zones of the different scenarios do not change considerably compared with the present state. However, the infiltration rates for the maximum scenario are orders of magnitude higher. The results further demonstrate the high fraction of Birs water (>60%) for the scenario with maximum opening (Fig. 5.7). With regard to groundwater quality the relative amount of surface water is an indicator for the potential pollution of the groundwater by infiltrating river water.

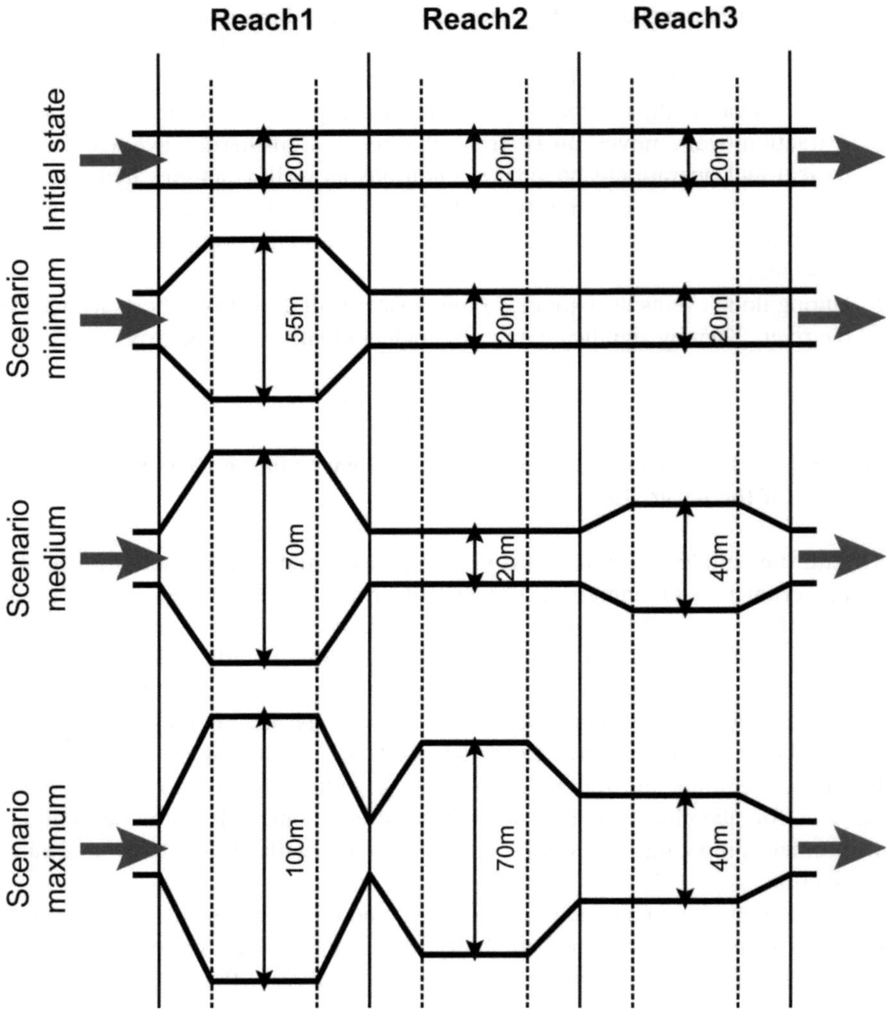

Fig. 5.5 Concept for restoration scenarios and river widening

5.1.6 Conclusions

A systematic consideration of groundwater protection in urban development and the implementation of groundwater management systems can serve as a decision tool for project planners and official departments. This allows ongoing adaptation dealing not only with current issues but also with future demands under changing conditions.

Adaptive management optimization strategies represent a coordinated iterative process of selecting and testing hypothesis of responses to management interventions. Risk assessment for water pollution, reservoir depletion, flood protection in

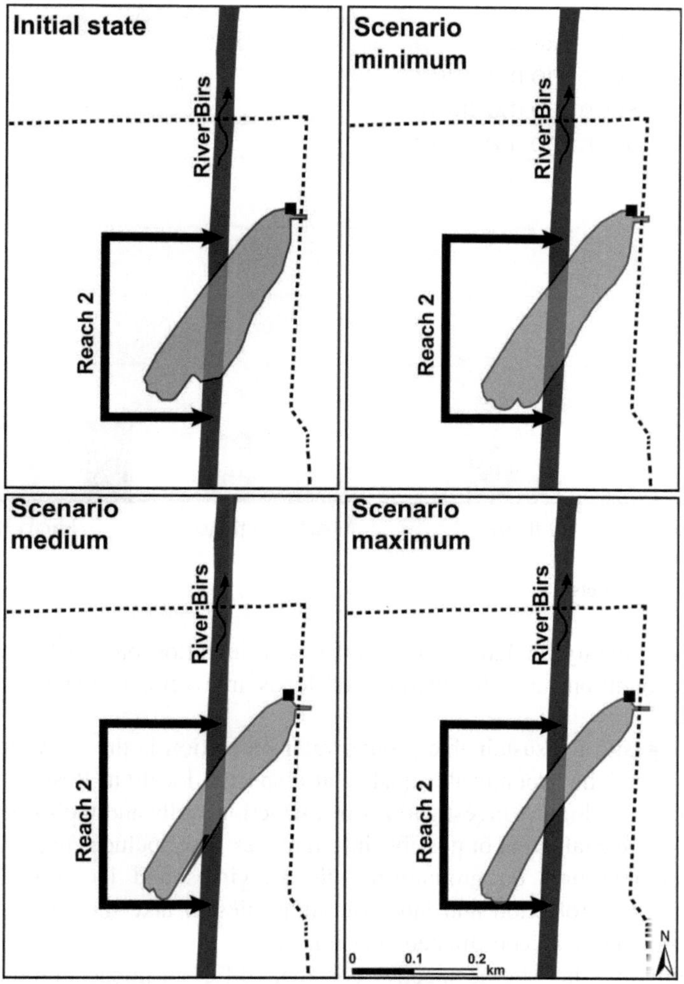

Fig. 5.6 Restoration scenarios

relation to climate change and river restoration, etc., has to be approached in regard to current and future use conflicts. In order to meet these goals, the combination of results from basic research focusing on process understanding and practical applications of efficient adaptive water management is necessary. These concepts also facilitate strategies for reorientation of current groundwater and surface water management practices. Successful application of the developed methods and forward-looking strategies are the scientific basis for decision making by the government and administration.

Extending current protection concepts with process-based approaches that consider the interaction between surface and subsurface waters could enhance sustainable development of groundwater resources. Knowledge of the composition of

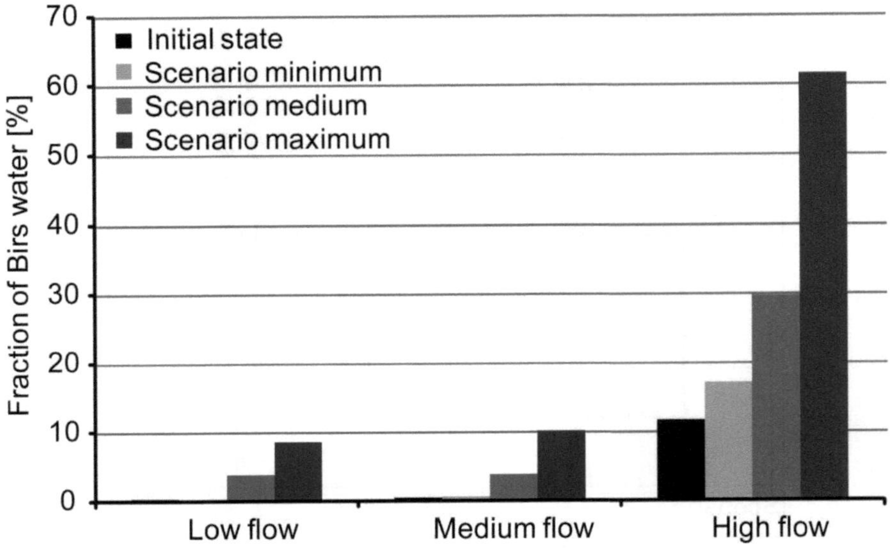

Fig. 5.7 Flow budgets

groundwater quality, including an adequate consideration of variable hydrologic boundary conditions and fluctuations of loads in rivers, is therefore of great importance.

A prerequisite for sustainable groundwater protection is the knowledge of the development of the groundwater quality at a specific location. Results from site-specific hydrogeological investigations of extraction wells and their operation are the basis for the evaluation of possible interferences. This includes impacts of water engineering measures on groundwater flow regimes and its usage, and also facilitates flood protection and land use authorities to take the various interests into account and make coordinated decisions.

Furthermore, a holistic perspective considers different impacts on the regional groundwater flow regime simultaneously, recognizing that impacts are not only taken as locally limited but could have effects on the regional scale. Therefore, the principal stresses on the system, such as groundwater extractions, injections, and recharge as well as water engineering measures and their impacts on the groundwater flow regime have to be taken into account.

Effective and efficient groundwater protection during flood events and water engineering measures along rivers demands detailed hydrological and geological knowledge as well as the willingness to suggest dynamic changes of hydrological conditions and load variations in rivers. By means of planning, organizational and technical measures the options for water engineering measures along rivers increase. The definition of goals helps to evaluate the impact of individual measures on a larger scale of the groundwater system.

The principle of resource protection is based on prevention and reduction, respectively, of contaminant release into the environment and on the conservation of groundwater resources. Some of the measures are related to the efforts to reestablish some of the natural function of riverine environments.

The changes in interactions between surface and subsurface water systems, when applying engineering measures along rivers often cannot be adequately evaluated based on existing groundwater protection concepts. Thus, the protection concepts could be extended by process-based approaches.

Other aspects are related to the multiple interests concerning water use and protection, which challenge the intentions of water engineering and groundwater protection schemes that can only be solved by simultaneously considering the different interests. However, for appropriate measures, such as reestablishment of river continuum, creation of retention zones, widening of river reaches, etc., the time schedule is far beyond in all of the river rehabilitation programs A complex system of competence assignment in the different countries makes a realization of specific projects time consuming, if not impossible.

It is generally accepted that the approaches for the management of water resources include processes at the catchment scale, the interactions between surface and subsurface waters, wetland processes, the state and development of terrestrial and aquatic ecosystems, as well as the consideration of issues concerning water quality and water budgets. The setup of appropriate groundwater monitoring systems, field experiments, and groundwater models allow the definition of specific groundwater system profiles. Based on these profiles and by applying scenario techniques, the essence of the dynamics of the groundwater flow regime and of capture zones of wells can be understood. The understanding of the processes will be a starting point for the optimization of resource management due to changing hydrological and operational boundary conditions.

5.2 Engineering Hydrogeology

Jannis Epting and Peter Huggenberger

Infrastructure development and associated alterations in land use often only consider the benefits for the improved infrastructure itself and planning largely takes the pragmatic form of engineering for short-term economic objectives. This often leads to adverse effects on subsurface resources and groundwater flow regimes.

Surface and groundwater monitoring during engineering projects usually is restricted in order to comply with existing laws and regulations governing water quality issues during construction activities. Some projects undertaken under outdated legal frameworks, i.e., some 30 years ago, would not be approved today because more restrictive laws pertaining to groundwater, as well as changed perceptions and policy concerning groundwater and its sustainable use, now apply.

Our concepts and methods for sustainable groundwater use in urban areas (Sect. 3.1) demonstrate that environmental impact assessments not only have to include above-ground impairments, such as ground motions with effects on existing buildings and infrastructures, as well as noise exposure and air pollution, but also the negative impacts on groundwater flow regimes. In this section, we discuss some strategies how to understand and predict the cumulative effects of the numerous single impacts to subsurface resources. Subsequently, we illustrate the approach for two case studies.

5.2.1 Impacts of Urban Infrastructure Development

Subsurface constructions can result in significant changes to subsurface resources and in groundwater quality and dynamics of both local and regional groundwater flow regimes. While some changes only temporarily affect urban systems during construction, others are permanent, like the reduction of cross-sectional groundwater flow and aquifer-storage capacities. Together with various sources of groundwater pollution observed in urban environments, subsurface construction may interfere with a previously balanced urban groundwater flow regime.

Impacts that temporarily affect urban subsurface systems during the construction include groundwater extractions and injection on the construction site as well as drawable sheet pile walls and slide pales. Impacts that will be permanent include parts of the construction extending below the groundwater table, permanent sheet pile walls, pales and cement injections for subsurface stabilization. Permanent impacts will change aquifer properties in a virtually irreversible way.

5.2.2 Concepts for Urban Infrastructure Development

When infrastructure projects that influence the subsurface and groundwater are realized several goals can be formulated. On the regional scale, general goals include the:

1. Minimization of changes to subsurface resources and groundwater flow regimes during and after construction, including the maintenance of the courses of regional and local groundwater divides, dimensions of groundwater budgets, and groundwater flow velocities.
2. Guarantee of the supply of groundwater (quantity and quality) for existing users and consideration of additional future groundwater use in the region.
3. Long-term improvement of subsurface resources and groundwater quality, with main focus on former industrial sites.

At the local scale, in the vicinity of the construction site, goals should focus on the:

1. Minimization of backwater effects behind parts of the construction extending below the groundwater table.

2. Prevention of the development of stagnating groundwater zones close to construction elements, extending below the groundwater table, these elements can act as barrier to groundwater flow and would reduce the storage volume of the aquifers.
3. Technical solutions guaranteeing safety standards on the construction site.

As the individual goals may interfere with each other and, together, may not necessarily lead to a desired overall goal, techniques that facilitate the comparison of interferences have to be applied. This can be accomplished by the development of scenarios and the implementation of equivalence and acceptance criteria (Bedford 1996). They assess the technical benefits of the different engineering projects, the supply situation for groundwater users, the development of the groundwater flow regime, and the improvement of overall groundwater quality. Together with equivalence and acceptance criteria, impacts on groundwater flow regimes and groundwater quality can be compared and evaluated Given the multitude of anthropogenic processes occurring in urban areas, it is difficult or even impossible to make comparative studies to a system state that describes a potentially uninfluenced natural environment. Therefore, for reference, the initial state before the beginning of major construction phase should be chosen.

Subsurface resource and groundwater management system can include monitoring systems for groundwater and river levels and quality as well as high-resolution numerical geological and hydrogeological models combined with scenario development (Sect. 4.2). Besides a simple documentation of changes in subsurface resources as well as groundwater quantity and quality, the goal of the management systems is to detect undesired developments in advance. In addition to long-term strategies in subsurface resource and groundwater monitoring, short-term monitoring programs can be setup during drawdown tests. With 3D geological and hydrogeological models possible scenarios can be developed, as:

1. Comparison of engineering projects.
2. Simulation of important project phases in advance.
3. Optimization of subsurface resource and groundwater management strategies.
4. Investigation of changing hydrological constraints.
5. Worst case scenarios.

5.2.3 Case Studies

Within the first case study, we illustrate how to accomplish groundwater management and protection during tunnel road construction (Epting et al. 2008a, b). In the second case study, we highlight groundwater protection during remedial measures of a subsided highway and a river dam, both constructed on gypsum containing rocks (Epting et al. 2009a, b).

5.2.3.1 Groundwater Management and Protection During Tunnel Road Construction

In this case study, we deal with the various stresses on a groundwater system during the construction of a tunnel highway in the northwestern area of Basel, Switzerland (Fig. 5.8). Parts of this area formerly were contaminated by industrial activities. Today, urban groundwater resources are extensively used by industry. During single construction phases, considerable groundwater drawdown was necessary, leading to significant changes in the groundwater flow regime. Furthermore, sufficient groundwater supply for industrial users and possible groundwater pollution due to interactions with contaminated areas had to be taken into account (Sect. 5.3).

The subsurface highway construction is 3.2 km long and connects the French highway A35 (Mulhouse – Basel) with the Swiss A2 (Basel – Gotthard – Milano). It is divided into four sections, of which about 87% are tunnel constructions; the remaining 13% consist of the bridge across the Rhine and the various tunnel entrances (Fig. 5.8). Construction started in 1994 and the whole highway project was completed by the end of 2008. The progressive shift of the construction sites, requiring different drainage systems, affected the groundwater flow regime throughout construction. Within the area, groundwater resources are extensively used by industry for processing or cooling (Fig. 5.8). A total of 13 industrial wells are operated in the vicinity of the construction site, the average amount of groundwater extracted from the aquifer is about 30 l s^{-1}, and approximately 3.5 l s^{-1} are injected back to the aquifer.

For all project phases, we evaluated changes of the groundwater system considering the various goals and requirements. We applied a groundwater management system, comprising extensive groundwater monitoring, high-resolution numerical groundwater modeling, and the development and evaluation of different scenarios. Selected examples focus on (1) a construction phase that is associated with considerable changes of the groundwater flow regime resulting in the turnaround of flow lines and shift of groundwater divides and (2) the evaluation of changes in local aquifer properties and groundwater flow regimes after construction.

During the construction of the subsurface freeway access and exit roads to the main tunnel, called "Tunnel Luzernerring" (Fig. 5.9) with groundwater extractions up to approximately 140 l s^{-1} (October 2003 to Mai 2007) a combination of open sump drainage and the dewatering of residual groundwater in areas enclosed was chosen. The exit road crosses below the main tunnel road and is thus the deepest part of the entire construction requiring maximum drawdown. In total, 26 extraction wells and three injection wells were installed to achieve the required drawdown of the groundwater table. Part of the extracted groundwater is injected back to the aquifer in three injection wells at distances 150 to 250 m from the construction site. The remaining amount of extracted groundwater is channeled to the Rhine.

The monitoring network comprised a total of 44 observation wells for continuous measurements of the hydraulic head. A total of 21 observation wells were

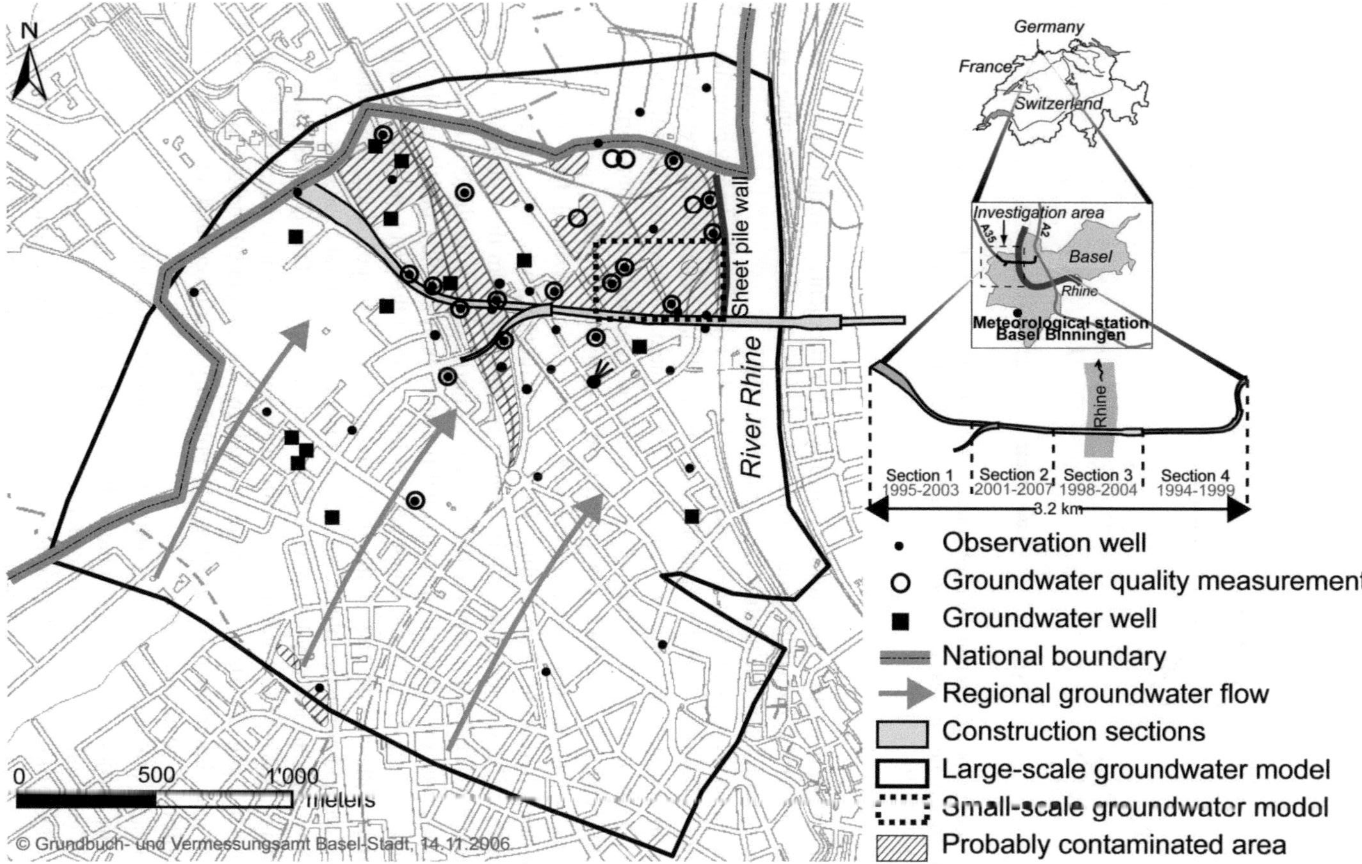

Fig. 5.8 Investigation area in Basel (northwestern Switzerland). Note that the highway tunnel runs at an angle of approximately 60° (counterclockwise) to the regional groundwater flow. In the *middle of the right hand column* the years are given within which each section of the construction was completed. The small-scale groundwater model is outlined in Sect. 4.4

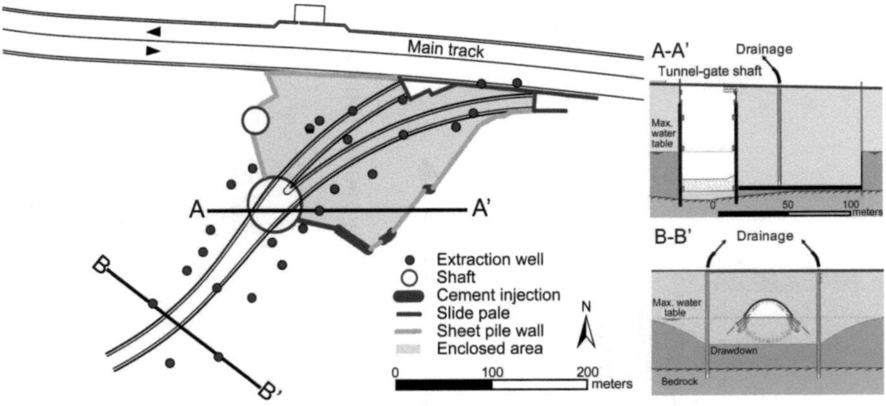

Fig. 5.9 Illustration of different drainage systems employed. Cross section A–A′: dewatering of residual groundwater in areas enclosed; cross section B–B′: open sump drainage for location see Fig. 5.8 section 1/2

sampled regularly for groundwater quality measurements. Furthermore, the extracted water for industrial groundwater use and for sediment settling tanks on the construction sites was sampled at regular intervals.

We set up a groundwater model which we adapted continuously, finally covering an area of 2,720 m × 2,860 m (about 8 km^2; Fig. 5.10). The spatial discretization resulted in cell sizes varying between 5 m × 5 m (near the construction site) and 30 m × 30 m in totally 132,500 cells. We chose an approach with four horizontal layers to vertically integrate the construction. Construction parts and cement injections we integrated either as inactive cells or as horizontal flow barriers with defined hydraulic permeability. During construction, we made progressive adjustments. The surface of the aquifer base (interpolated from the information of more than 400 boreholes), and the distribution of horizontal hydraulic conductivity zones was based on different type and quality data sets available from the geological database (Sect. 4.1). Model boundary conditions are of the first type (fixed head) along the southern side and of the third mixed type (leakage) along the Rhine. The western and northern boundaries were considered as general head. For groundwater flow simulations, the 3D finite difference code MODFLOW (Harbaugh et al. 2000) was employed in combination with the graphical user interface Processing Modflow (Chiang 2005). As a routine procedure, we calibrated the groundwater model at least biannually, by updating the boundary conditions and adjusting the permeability of sheet pile walls.

All results are being compared with the calibrated initial state in March 2003 before the major construction phase. Furthermore, we compared the various model calculations and developed scenarios and validated them by means of groundwater budgets through defined regions, the course of well capture zones as well as the description of simulated hydraulic heads, flow paths, and velocities (particle tracking).

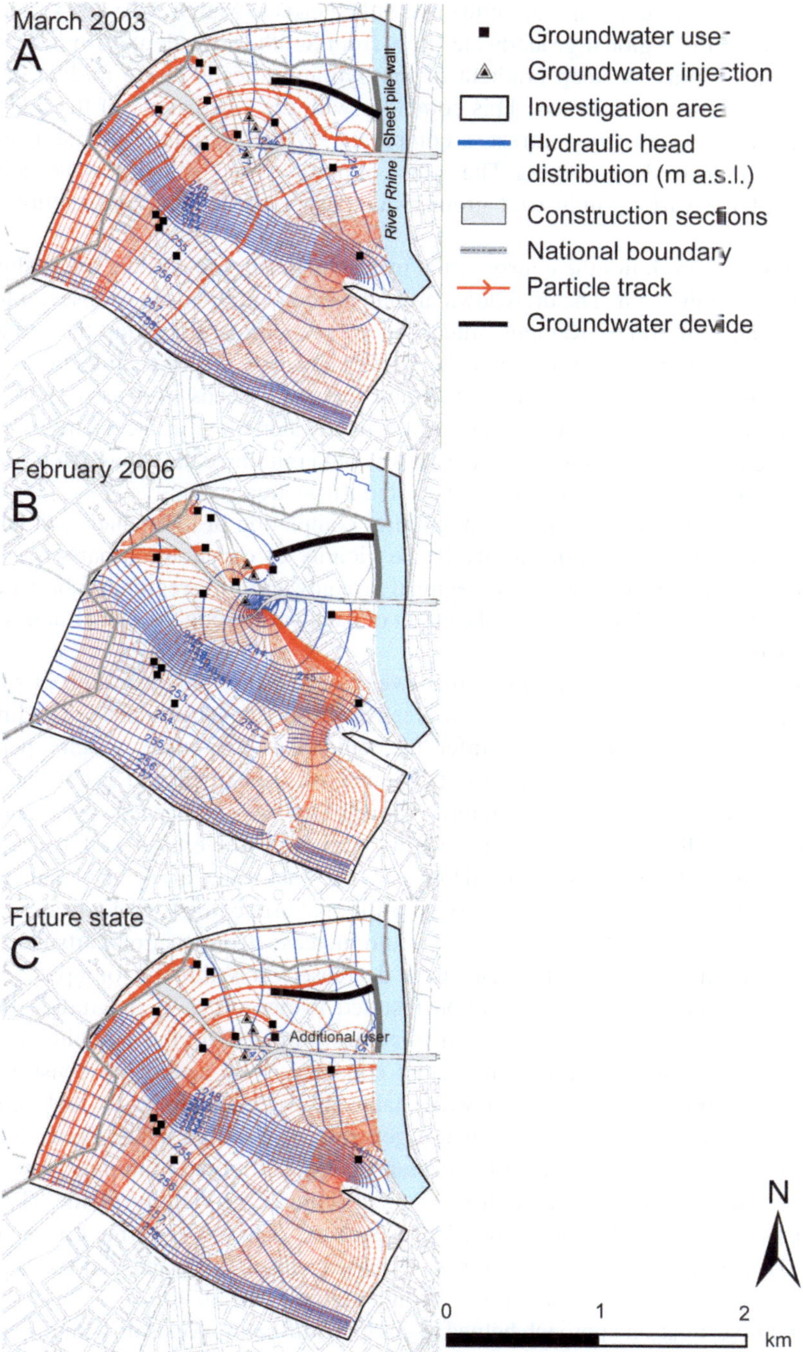

Fig. 5.10 Visualization of hydraulic head distributions (0.5 m resolution) and flow paths illustrated by particle tracks (distance between *two arrow heads* indicates 50-day travel time) for three modeled situations: (**a**) March 2003, (**b**) February 2006, and (**c**) Future state. Whereas the results for **a** and **b** derive from model calibrations, those for **c** are based on a simulated scenario

The calculated contour map of the hydraulic heads in March 2003 shows a main direction of the regional groundwater flow from South to North and from West to East (Fig. 5.10a). A steep gradient of the hydraulic heads in the middle of the model area can be observed. This coincides with the steep slope of the bedrock surface in this area. By contrast, a comparatively low hydraulic gradient in the northern industrial area occurs. The course of particle tracks illustrates the capture zones of the various industrial groundwater users and of the construction site drainage.

Figure 5.10b shows the course of the hydraulic heads and particle tracks during maximum drawdown. The drawdown around the construction site is observable up to a distance of 500 m. Regions influenced are predominantly within the southern model area. The distribution and the performance of the three injection wells were confirmed. Our modeling results and hydraulic heads measured show that the supply of groundwater for industrial groundwater users in the vicinity of the construction site was assured. Moreover, we could minimize the change of the local groundwater flow regime of the northern industrial area.

Figure 5.10c illustrates the situation after completion of the tunnel road taking into account the final permeability in the vicinity of construction parts reaching below the groundwater table. The regional groundwater flow regime is comparable to that of March 2003. However, the effect of backwater along the main track is still observable.

After completion of the tunnel road we reviewed the long-term impact of the construction measures on the groundwater system and the performance of technical measures. The potential of the aquifer that could be used (storage capacity, permeability) is decreased in the vicinity of the tunnel road construction. The connectivity of the groundwater and minimum permeability was enhanced by technical measures, such as the installation of highly permeable culverts as well as drawing sheet pile walls and slide pales. However, the backwater along the main track ranges between 0.1 and 0.4 m. In case an additional user extracts groundwater north of the main track the backwater effect could increase. Construction parts reaching below the groundwater table lead to a decreased flow-through of up to 51%. Pumping tests and model calibration resulted in one order of magnitude lower permeability in vicinity of the construction compared to the situation before construction. This decrease in hydraulic conductivity is caused by construction parts reaching below the groundwater table as well as pile walls and cement injections. Additional groundwater use generally is possible in the South; in the North careful clarifications are necessary. An additional pumping test was performed to evaluate extraction rates for a fountain in vicinity of the "Tunnel Luzernerring" named "Wasserspiel Vogesenplatz." The test confirmed once again that the permeability of the aquifer is decreased substantially and that the desired extraction rates of more than 20 l s^{-1} could not be realized.

Our investigation approach helped to compare different proposals with respect to feasibility and impact on the groundwater flow regime during construction and after completion of the tunnel road. We could simulate important project phases in advance and compare different groundwater management strategies. This helped us evaluating the localization and operation of extraction and particularly injection wells; localization and dimension of culverts, sheet pile walls, and slide pales; localization of additional observation wells; and prediction of additional groundwater use in the future.

The results show that due to the groundwater management, changes in the groundwater flow regime, especially towards the North, were comparatively low. With the aid of groundwater modeling, the dynamics of the groundwater flow regime under changing spatial and temporal constraints could be simulated and evaluated. Next to the management of the various groundwater extractions and injections, the requirements for groundwater protection (groundwater flow regime, groundwater quality) were achieved satisfactorily. Our groundwater management system also helped to identify changes in groundwater chemistry. Negative consequences for the industrial groundwater users could be minimized. It was not necessary to install supplementary injection or interception wells to ensure the supply of groundwater for the industrial users, or to prevent the attraction of contaminated groundwater.

5.2.3.2 Groundwater Protection During Road Remediation

Infrastructures that are constructed on unstable geologic formations are prone to subsidence. While the characterization and modeling of flow in heterogeneous and fractured media has been investigated intensively, there are no well-developed long-term hydrogeological research sites for gypsum karst. Additionally, systems for monitoring the evolution of karst phenomena are rare. In this case study, we document the integration of different investigative methods in the context of an engineering project for the upgrade of a subsided highway located beside a river dam.

The project area is located in the Lower Birs Valley southeast of Basel (Fig. 5.11). Over the last 30 years, subsidence of a man-made river dam and an adjacent highway has been observed. Surface–groundwater interaction is dominated by the hydraulic river head and variations in river bed conductance upstream of the dam during flood events. Upstream of the dam, river water infiltrates into the highly permeable fluvial gravels and the weathered bedrock follows the hydraulic gradient around and beneath the dam and exfiltrates downstream into the river. These processes and the drilling of several boreholes in the 1990s that likely left a stratigraphic connection and hydraulically connected aquifers have led to the karstification in the soluble units of the “Gipskeuper” and resulted in an extended weathering zone within the bedrock as well as in the development of preferential flow within voids and conduits.

To prevent further subsidence, construction measures were carried out in two major project phases in 2006 and 2007. The highway was supported by 166 piles and by a sealing pile wall, consisting of approximately 300 piles (Fig. 5.11), to prevent infiltrating river water from circulating around the dam and beneath the foundation of the highway. Piles extend down to the nonweathered rock at a depth

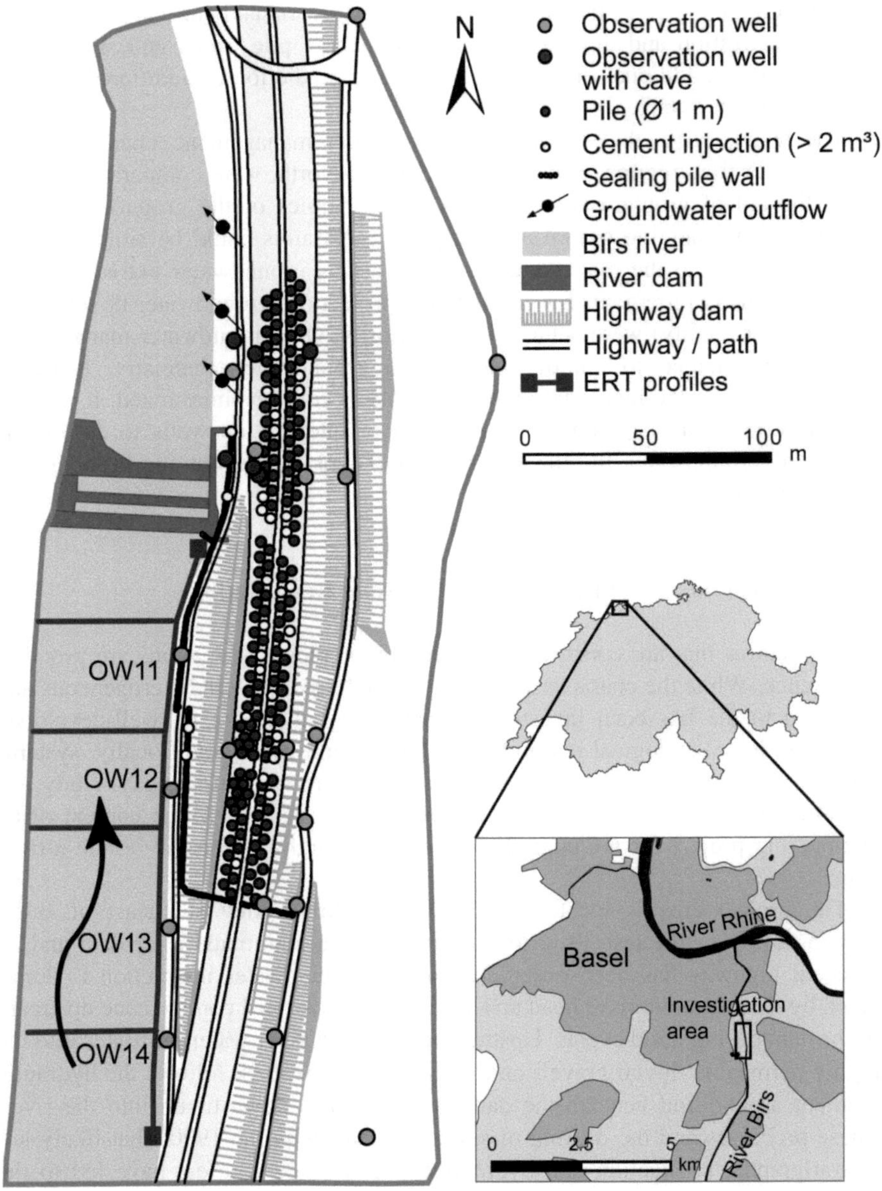

Fig. 5.11 Investigation area in the Lower Birs Valley southeast of Basel illustrated together with the applied hydrogeophysical investigations (see Sect. 4.3) and engineering measures

of 20 to 25 m. Caves encountered when the piles were being constructed were filled with a total of 168.2 m^3 of supplementary cement, in order to plug underground water channels and stabilize the ground beneath.

In the current case study, sporadic measurements revealed that subsidence of the highway and the river dam has increased rapidly over the last 10 years. At the beginning of the project, system knowledge was limited to purely conceptual models and sparse accurate monitoring data. As regards the locally specific engineering problem, and in order to plan appropriate remedial measures, we recommended setting up instruments that allow this complex system to be examined under various hydraulic conditions. This included an assessment of the current groundwater flow regime as well as the subsidence mechanism and its development over time. Such approaches require, along with the installation of surface and groundwater monitoring systems, specific field campaigns and modeling strategies to investigate the relevant processes. Furthermore, the developed tools should have not only a monitoring character, but they should also enable long-term solidly based predictions to be made concerning the future evolution of the system and potential subsidence.

In compliance with existing regulations, an observation network was installed in order to monitor surface and groundwater quality in the vicinity of the construction site and regional drinking water supplies further downstream. Additionally, the observations allowed us to identify and evaluate the relevant processes. Online monitoring allowed us to early detect and document changes in hydraulic constraints in the vicinity of the construction site and the dam. In the event of breakthrough events and the mobilization of void fillings and cement injections, we could have initiated interventions to prevent surface and groundwater pollution.

Since 2006 we investigated changes in the groundwater flow regime continuously by different methods that also allow the evaluation of the long-term performance of the infrastructures. Geological (outcrops, lithostratigraphic information of boreholes), hydrometrical (extensive groundwater monitoring, dye tracer tests), and hydrogeophysical (Electrical Resistivity Tomography, ERT) field data of varying quality were integrated into high-resolution 3D hydrogeological (Fig. 5.12) and 2D karst evolution models (see Sect. 5.4). We validated the applied investigative methods and determined the sensitivity of relevant parameters governing the processes.

Boreholes provided data and general lithostratigraphic information, including details about the vertical extension of the weathered Gipskeuper and significant permeable zones, as well as already developed voids. Ever since the 1990s, the drilling of several boreholes has left a stratigraphic connection and locally stimulated karstification. Initial, coarse cross sections could be developed using the information from the national geological map. Additional data from lithostratigraphic information was obtained from the reports made during the installation of the piles, resulting in more precise cross sections of the investigation area. We could confirm hydraulic links within the investigation area by a dye tracer test with groundwater flow velocities typical for conduit systems. These results indicated that the karst system is already well developed, whereas solution conduits developed along a system of fractures and interconnected joints, suggesting a 3D conduit network.

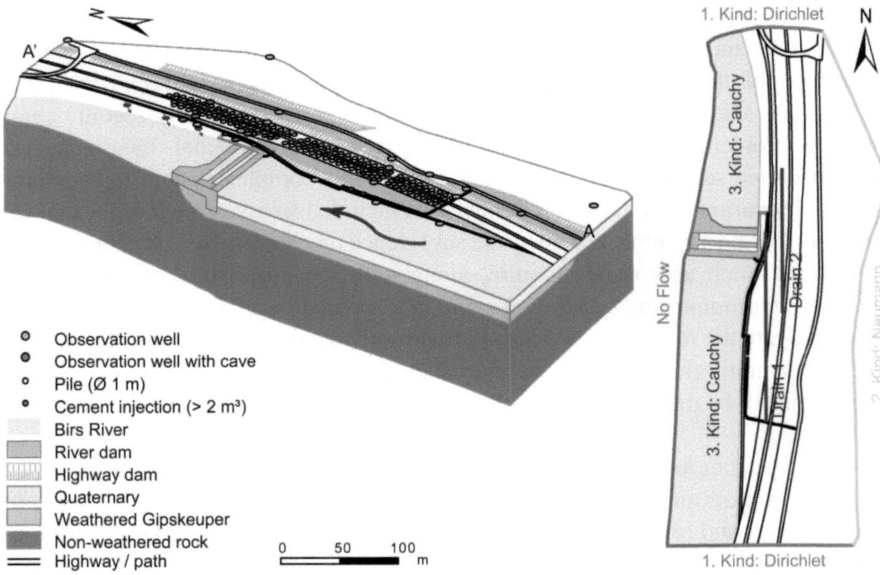

Fig. 5.12 *Left*: 3D geological model. *Right*: conceptual 3D groundwater model setup with hydraulic boundary conditions

Results from surface and underwater ERT measurements (see also Sect. 4.3), taken at different hydrologic and geotechnical boundary conditions, both before and after the construction measures, provided data and allowed the description of:

1. Preferential flow in the shallow subsurface.
2. Zones that are related to groundwater flow around the dam, including flow dynamics.
3. Zones that are related to groundwater flow beneath the dam.
4. Drainage phenomena of karst features such as voids and conduits.
5. The weathering horizon within the Gipskeuper.
6. Near-surface faults and fracture zones.
7. Buried paleochannels.
8. Sediment thickness followed by weathered zones beneath the river upstream of the dam.

Due to the multiple data sources of varying quality and hydraulic data from high-resolution 3D hydrogeologic model, it was possible to partially eliminate ambiguity in data interpretation and to describe the relationship between the different observed features in a spatial context.

Figure 5.13 illustrates the simulated groundwater flow regime at average discharge before and after the construction measures. The groundwater flow regimes for both model scenarios clearly show the influence of the dam structure and the effect of the drains. The gradient in the nonweathered rock is steeper than in the weathered Gipskeuper and the Quaternary cover. For the simulation after

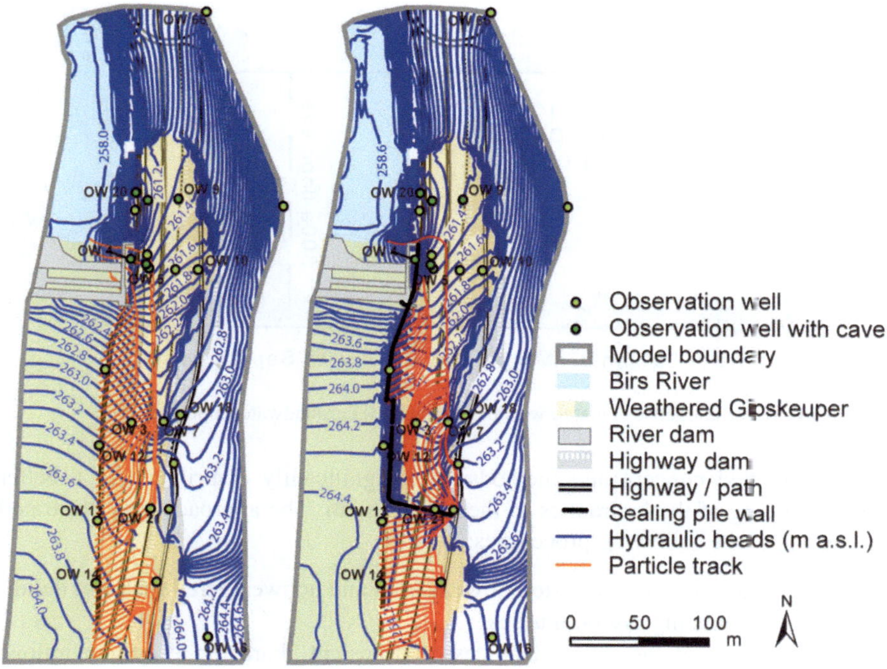

Fig. 5.13 Visualization of hydraulic heads and particle tracks (0.1 m resolution). *Left*: Groundwater flow regime at average river discharge before construction measures (07.02.06). *Right*: Groundwater flow regime at average river discharge after construction measures (15.01.08)

the construction measures, the sealing pile wall was integrated. As an effect of the sealing pile wall, backwater can be observed towards the river.

In Fig. 5.14 water budgets for the general head boundary up- and downstream of the dam are illustrated together with model outflow through the drains. Model inflow is dominated by the general head boundary upstream, indicating river water infiltration; model outflow is dominated by the general head boundary downstream, an indication of groundwater exfiltration (an approximate measure for the diffuse component of flow) and through the drains (approximate measures for the conduit component of model outflow). Model inflow (general head boundary upstream) generally equals model outflow (general head boundary downstream and drains). The figure also shows the relative contributions to the flow systems (diffuse and conduit model outflow; Drains 1 and 2) before and after the major flood event. The data illustrate that in general during flood events, the relative contributions of the conduit component of model outflow increases. After moderate- to medium-scale flood events, the relative contributions of flow components return to ratios observed before the events. However, major episodic flood events, as the 300-year flood of 9 August 2007 (Fig. 5.15), can considerably change the relative amounts of flow components also after the event.

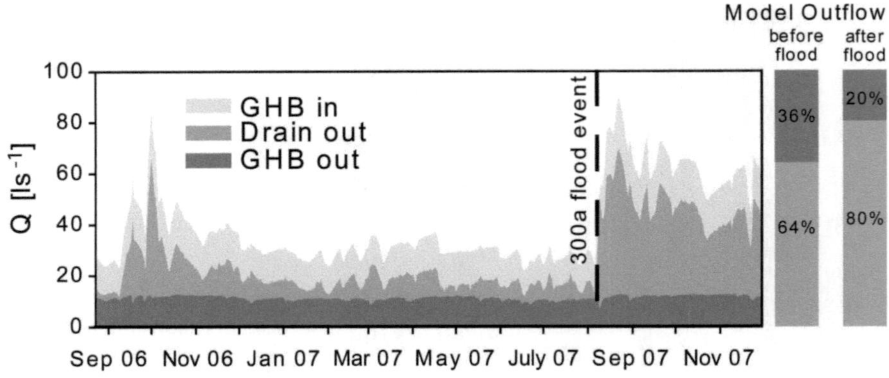

Fig. 5.14 Water budgets calculated with the transient 3D groundwater model

The applied concept and methods have significantly contributed to a better understanding of the hydraulics of the karst system. The approach was illustrated by means of the following procedures:

1. Determination of the extension of weathered and nonweathered rock and definition of preexisting discontinuities.
2. Identification of the relevant processes (transient character of system inflow, description of slow and fast flow components).
3. Evaluation of the influence of episodic major flood events, accompanied by the flushing of conduit fillings and the inflow of undersaturated water.
4. Investigations of the long-term development of the system.

Comprehensive studies of transient 3D hydrogeologic model facilitated the evaluation of the relevant groundwater hydraulics and revealed the dynamic character of the groundwater flow regime during low frequency flood events, including river infiltration and diffuse and conduit components of model outflow. 3D hydrogeologic model facilitates current state descriptions of karst systems and provides sufficient information for:

1. Estimating the transient composition of water budgets.
2. Describing the transient character of the groundwater flow regime.

Fig. 5.15 300-year flood of 9 Aug 2007

3. Simulating and evaluating short-term impacts on processes such as those which occur during episodic flood events.

However, we could only achieve long-term evolution of karst systems and prognosis by setting up models that account for the change in hydraulic properties with time. The results we present in Sect. 5.4 show how 2D karst evolution modeling can be calibrated to describe the current state of karst systems, and can be used for prognosis of system development and subsidence risk assessment.

We could demonstrate that the applied methods for karst aquifer characterization complement each other and allow the interpretation of short-term impacts and long-term development on system dynamics in the context of groundwater flow regimes of karst areas. This includes the description of the transient character of the groundwater flow regime during and after episodic flood events (surface–groundwater interaction, conduit and diffuse model outflow) as well as the evaluation of time scales for karst evolution. Results allow the optimization of investigative methods for similar subsidence problems, leading from general measurements and monitoring technologies to tools with predictive character.

5.2.4 Conclusions

Integrated conceptual approaches incorporating methods for adaptive groundwater management system can help to meet challenges posed by major constructions in sensitive urban environments. In order to avoid a permanent negative impact to groundwater flow regimes, particularly concerning quantitative and qualitative groundwater protection and irreversible deterioration of aquifer systems, recommendations for the optimization of the groundwater management should be proposed and constructional arrangements provided. Urban constructions allow comprehensive hydrogeological investigations that otherwise would not be possible.

5.3 Contaminated Sites in Urban Areas

Peter Huggenberger, Jannis Epting, Eric Zechner, and Annette Affolter

In urbanized and industrialized regions, water resources, and particularly groundwater, are subject to many pollution risks related to different types of ancient industrial activities (United Nations for Environment Program/Agency for the Environment and Energy Management 2005; European Environmental Agency 2006). Because of the spatial extent of GWB in these regions, many point or diffuse pollution sources may need to be considered during the development of new building and transport infrastructures.

In Europe large areas in urban agglomerations have been industrialized in the nineteenth and twentieth centuries. Since then, some of the industrial activities lead to pollution of the subsurface, depending on the produced goods or on the history of accidental release of contaminants due to chemical spills or hazardous fires. In addition, there are many other sources of pollution that are distributed over the entire urban area, including their agglomerations. In fact, almost each urban agglomeration has to deal with hundreds or thousands of contaminated sites, including diffuse or point sources.

Many of these former industrial sites are now being transformed into modern business centers or residential areas. Furthermore, in urban areas industry production was often settled along national borders. The "Trinational Euro District Basel" is a good example on how the former industrial areas along the national borders are becoming increasingly valuable land resources for future urban development. At the same time these zones are located in areas with valuable groundwater resources, presently used for different industrial processes and cooling.

Potentially or actually polluted sites are registered and evaluated according to the federal soil protection and contamination ordinance. Cleanup of contaminated sites is generally organized according to risks and specific selection criteria that vary in the different countries. After a huge effort to clean-up several polluted sites in the eighties, we now observe a policy change, away from clean-up towards monitoring and natural attenuation. An exception represents some large hazardous sites.

The construction activities during urban development can impact regional groundwater flow regimes and new constructions (subsurface traffic lines or large buildings) may affect urban groundwater systems temporarily during construction as well as permanently after completion. Potential impacts can include a reduction of the cross-section for groundwater flow and a decrease of aquifer storage volume. Some of these impacts are permanent, while others, such as construction site drainages, only affect the groundwater flow regime temporarily during the construction period. In many large infrastructure or building projects a lowering of the groundwater table during construction is required. Due to the pumping, groundwater flow regimes change. This can lead to remobilization of contaminants from ancient polluted sites.

As a result of the numerous subsurface construction projects in the last years and the conversion of ancient industrial sites into business or industrial research centers there is need for action and a concerted effort during subsurface planning. This includes historical and technical investigation as well as risk assessment and the evaluation of monitoring and remediation strategies of contaminated sites.

The urban area of Basel is an excellent example to illustrate these developments and strategies in handling these problems during the development of urban infrastructure. Since the region turned into a major industrial center for the chemical and pharmaceutical industry in the nineteenth century, vast areas have been or are likely to be contaminated. Contaminations mainly include residues and solvents from the color industry, such as BTEX, volatile organic compounds, chlorinated compounds and their metabolites, as well as metals. In addition, there are abandoned sites of small enterprises and sites with municipal or multicomponent waste (fillings of former gravel pits) on adjacent French and German territory. The nonproductive areas of the two countries are becoming increasingly valuable resources for future urban development. The aquifer is presently used in these areas by numerous municipal and industrial water suppliers. Due to different activities the groundwater flow regime changed several times, both at the local and the regional scale. The different activities are mainly due to the construction of infrastructure in the urban space.

As a conclusion, subsurface traffic- and service-lines and the construction of large complexes in the subsurface are some of the important drivers for changing groundwater flow regimes. As a consequence considerable risk with regard to mobilization of contaminants will thus be caused by changes to groundwater flow regimes. Two case studies allow to document these processes and to illustrate strategies for the construction phases.

5.3.1 Institutional Aspects of Cooperation in a Multinational Urban Context

Transboundary cooperation in environmental issues in the Basel area has made considerable progress. Primarily because of the activities of local groups, such as the "Trinational Euro District Basel," whose aim is to cooperate in the three countries in the domain of urban planning, healthcare, and traffic infrastructure, as well as the "Regio Basiliensis" (initiation and cooperation in transboundary projects) and the support of international organizations, such as the "Council of Europe," the Upper Rhine Valley and other European border regions have succeeded in voicing their interests in a fairly cohesive manner. However, concerning environmental issues, the continued emphasis of national governments on sovereignty and national interests has prevented international border regions from achieving such basic goals as infrastructure integration and harmonization of environmental policy.

There are several institutions which are in charge of the transboundary cooperation in the Upper Rhine area in the domain of groundwater encompassing politicians and environmental scientists. An important board is the groundwater expert panel of the French–German–Swiss conference of the Upper Rhine. This board supported the several initiatives of INTERREG projects in the groundwater domain ranging from the transboundary compilation of hydrogeological data (Wagner et al. 2001) to the development of tools that allow to predict the groundwater quality with respect to changes in agriculture policy in the three countries (i.e., INTERREG III A-Projekt MoNit 2006). The legal aspects are still matter of the three countries, which requires a certain effort of harmonization.

5.3.2 Case Studies

The two case studies of urban and transboundary groundwater projects we present are located in northwestern Switzerland and extend to both Germany and France

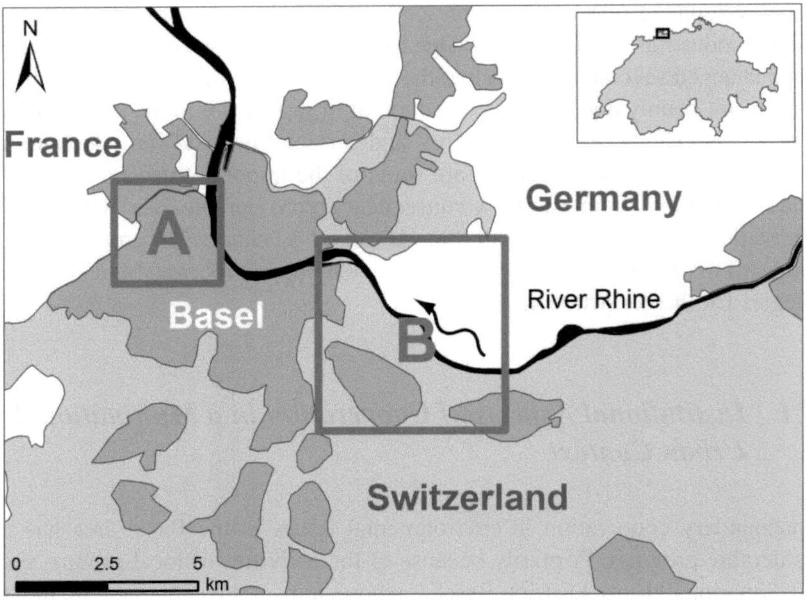

Fig. 5.16 Location of case studies. A: Groundwater management during highway construction; B: Ancient contaminated industrial sites

(Fig. 5.16). In both areas the shallow unconfined upper aquifer mainly consists of gravels of the lower terrace (late Pleistocene) deposited by the Rhine. The gravel depositions are intercalated with fine-grained flood-plain sediments that result in variable permeability within the aquifer. The thickness of the aquifer ranges between 15 and 35 m and is underlain by an aquiclude consisting of mud to clay rich sediments of Oligocene age. In the area, as illustrated in the second case study, the aquifer system is more complex due to a lower karst aquifer.

River–groundwater interactions along the Rhine are an important element in the regional groundwater flow regime. The groundwater table fluctuates phase-delayed and with reduced amplitude in response to the river level fluctuations of the Rhine. Depending on the hydrological constraints, the river can be a source or sink for groundwater.

To define the specific profiles of groundwater systems, we applied high-resolution groundwater models which we calibrated transiently against groundwater head data, river stages, and extraction and recharge rates. In examples we present, the strongly transient character of groundwater flow regime and/or river–groundwater interactions in urban areas is illustrated. We developed scenarios to assess the consequences of decisions and to optimize groundwater management.

In the first case study, we discuss strategies to understand and predict the cumulative effects of the numerous single impacts on groundwater resources during a major suburban development project (see also Sect. 5.2). Here we focus on a construction phase that was associated with considerable changes in groundwater flow regimes resulting in the reversal of flow lines and a shift of groundwater divides.

In the second case study, we analyze the development of groundwater pollution during the last decades in a heavily industrialized groundwater protection area. This includes the illustration of long-term changes to a groundwater body due to changed hydraulic boundary conditions.

5.3.2.1 Groundwater Management During Highway Construction

The highway construction that we outline as the first example is located in the northern part of Basel on the western bank of the river Rhine (details in Sect. 5.2). The project was realized step wise over a 10-year time period. Traffic lines and groundwater pumping including groundwater quality issues for different users had to be maintained over the entire construction period. The specific construction strategies for the various construction phases underwent progressive adaptations. These complex external boundary conditions required the setup of an instrument for monitoring and managing complex changes of the groundwater system for the different construction phases. The accuracy of the prediction of the groundwater model benefits from a total of 44 observation wells instrumented with automated water-level loggers that continuously measured the hydraulic head and temperature. Groundwater samples were regularly taken from 21 observation wells for monitoring a series of quality parameters. Furthermore, the quality control included analyses of the groundwater of the industrial groundwater users and to the sediment settling tanks on the construction sites itself at regular intervals.

During the entire construction period, the groundwater flow regime was affected by step wise progression of the construction sites requiring an adaption of the focus of the groundwater monitoring and management system. Depending on the mining or excavation techniques, the degree of complexity for groundwater drainage varied. The principal strategies were open or closed groundwater dewatering systems of the construction sites. These were either realized as open sump drainage, the dewatering of residual groundwater in areas enclosed by sheet pile walls, or a combination of both. The groundwater extracted from the construction site was generally discharged into surface waters, into the sewage system or recharged back to the groundwater. In all cases, the discharged water had to satisfy the specific quality standards.

A considerable change in the local groundwater flow regime in the northern industrial area was expected due to the open sump drainage in 2006. With the known polluted sites on both sides of the national borders of France and Switzerland, in the vicinity of the construction site, there was a considerable risk for mobilization of contaminants resulting from groundwater extractions and drawdown of the groundwater table. Contaminated areas may have suddenly lain in the capture zones of the industrial groundwater users or within the groundwater drainage areas on the construction site. As a consequence, positions of groundwater recharge and required recharge rates were evaluated in order to maintain the groundwater flow regime to the North and to minimize remobilization or deviation of contaminated groundwater (Fig. 5.17).

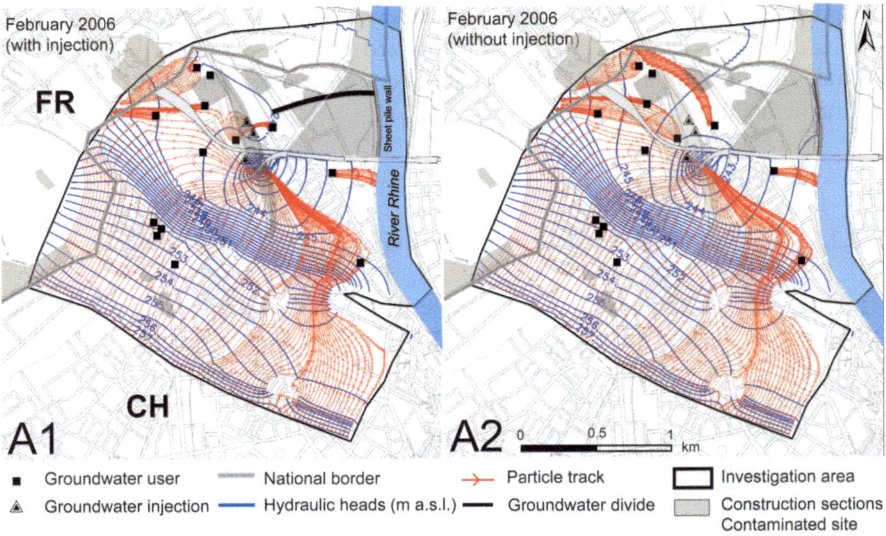

Fig. 5.17 Groundwater flow regimes for the first case study. A1: with groundwater injection. A2 without groundwater injection

Figure 5.17A1 shows the course of the hydraulic heads and particle tracks during maximum drawdown. A drawdown around the construction site was observed up to a distance of 500 m. Capture zones of pumping wells extended predominantly to the southern model area. The predicted position of the groundwater recharge wells and the positive influence on the stabilization of the groundwater flow regime in the areas of the ancient polluted sites could be confirmed during the entire construction period. Our modeling results and hydraulic heads measured in the monitoring system showed that the supply of groundwater for industrial groundwater users in the vicinity of the construction site was assured.

Figure 5.17A2 simulates a scenario of a failure of artificial groundwater recharge in the three injection wells. Due to such a failure the entire groundwater flow regime north of the steep slope would be changed. In such a case, the drawdown would be observed beyond the French border. The capture zone of the drainage would widen significantly and extend from the Rhine to the western model boundary. The groundwater divide would be displaced far to the North beyond the model boundary.

In most cases, the inflow of contaminated groundwater could be prevented. An exception occurred during the drainage of section 4 of the construction (see Sect. 5.2). In this case, groundwater modeling allowed to describe the transport flow patterns and to localize the origin of the contamination. Long-term monitoring will show if the recovery of the groundwater flow regime will reach preconstruction conditions.

Our management system was continuously adapted, based on the progress achieved on the various construction sections or risk profile in the case of old contaminated sites. Our interpretation of changes observed in groundwater quality measurements together with the modeling results allowed for an optimal localization of new observation wells. The extended knowledge of groundwater flow regimes could lead to a reduction and minimization of negative effects during the various construction phases and result in a sustainable development concerning use and management of the groundwater resource.

5.3.2.2 Ancient Contaminated Industrial Sites

In the area of the second case study we present, important drinking water production is located in the vicinity of several contaminated sites (Auckenthaler et al. 2010; Huggenberger et al. 2010). Aware of the risks, and due to measurable pollution of different origin and from different time periods, the drinking water production was protected since years by the infiltration of filtrated Rhine water, creating a hydraulic gradient towards the known contaminated sites and towards infiltrating surface waters from the river Rhine.

A total of 72 wells, 30 of them used for industrial use and 42 as drinking water wells, were integrated into a 3D numerical groundwater model. Subsurface parameters such as hydraulic conductivities and riverbed leakage factors were calibrated with continuous records of hydraulic heads from 121 observations wells. Our model was used to determine the capture zones of drinking water production wells and downstream flow from the most important polluted sites for different time periods with changing boundary conditions. Figure 5.18 shows the development of the groundwater flow regime before construction of a dam for a hydropower plant in the river Rhine and before the development of the water supply (B1), after construction of the river dam for the hydropower plant in the Rhine, which increases hydraulic gradients parallel to the Rhine (B2) and after the development of a municipal water supply in the mid-twentieth century (B3). Both the construction of the dam and the water supply lead to a significant change of the groundwater flow regime. For example, the direction of the downstream flow from the contaminated sites in the center of the investigated area changed from NNE (before 1954) to NW (between 1954 and 1956) and to E (after 1956). It is likely that the contaminants have been spread according to the predominant flow field at that time and that parts of them were retained in the subsurface according to the compound specific physicochemical conditions, the pore structures, and the mineral surfaces. This process would actually explain the large spreading of contaminants still measurable at many different locations at relatively low concentrations but with fluctuations around the detection limits for the required quality standards of the drinking water guidelines. In addition, our models illustrate the transboundary character of the different groundwater flow regimes over the last 50 years. It also clearly documents the fact that in future a change of production of one groundwater user would likely lead to changes in water quality for other groundwater users. As a consequence, groundwater management in this area has to include all the groundwater users not only on the left Swiss side of the river Rhine but across the national borders to Germany.

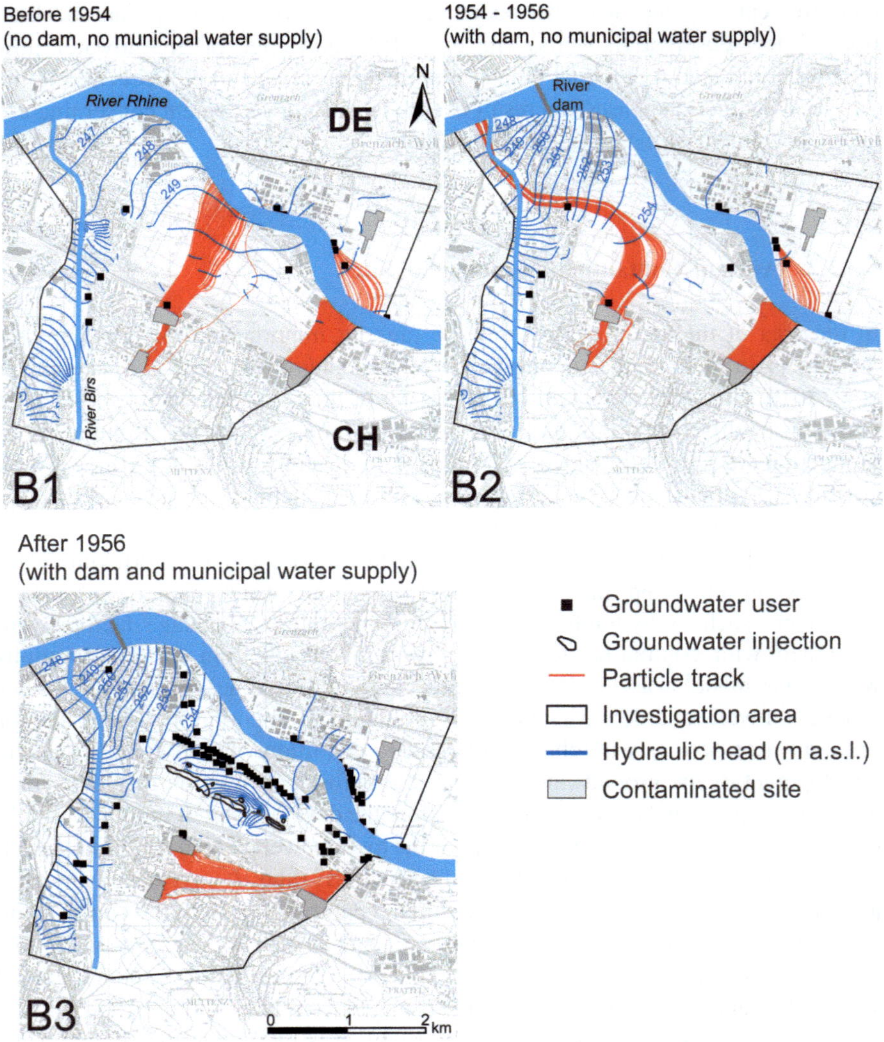

Fig. 5.18 Groundwater flow regimes for the second case study. B1: Situation before construction of the river dam. B2: Situation with river dam and before installation of the water supply. B3: Situation with river dam and water supply. Topographic cards are reproduced with permission of swisstopo (BA110446)

5.3.3 Conclusions

In order to predict, mitigate, or prevent environmental problems across national and legislative borders and to assure the supply of groundwater for municipal and industrial users, we choose integrated adaptive groundwater management approaches (see Sect. 3.1). Extending current protection concepts with process-based approaches as shown in the examples earlier could enhance the sustainable development of groundwater resources. Knowledge of the composition of groundwater quality, including the consideration of variable hydrologic boundary conditions and fluctuations of contaminant loads in rivers, is therefore of great importance. Key factors in investigating contaminant transport are the relevant boundary conditions as well as their development and origin, in particularly the depth and nature, of relevant substances. In the two case studies, the boundary conditions are dominated by the intense groundwater pumping and the rates of artificial recharge, a fact which might be typical for many urban groundwater management systems. Our modeling results also indicate that flow paths and groundwater velocities can vary considerably due to changes in hydraulic boundary conditions.

Management strategies for groundwater which are based on transboundary numerical simulations are still confronted with enormous implementation barriers. Confidence in their success is often low, and conventional and more expensive approaches, like extensive drillings and analytical programs, are preferred. The illustrated approaches can help to meet challenges posed in a sensitive transboundary urban environment. This includes the evaluation of contaminated sites, risk assessment for waste disposal, as well as the parameterization of numerical groundwater models. While some of this work may be specific to these case studies, it is expected that our overall conceptual approach and the methodologies will be directly transferable to other urban and transboundary areas. Thus, this is one step towards the application of new findings to complex practical problems.

As environmental problems generally do not stop at national boundaries, the exchange of information about major impacts to the groundwater flow regime requires communication between the neighboring countries.

5.4 Karst in Urban Areas

Jannis Epting, Peter Huggenberger, Eric Zechner, Daniel Gechter,
Ali Zidane, Markus Konz, and Douchko Romanov

Groundwater circulation in evaporite bearing horizons and the resulting evolution of karst frequently causes geotechnical problems such as land-subsidence or collapses. Such processes are of particular concern in urban areas where soluble geological formations coincide with vulnerable infrastructures as transportation systems. One of the most famous cities where infrastructures are prone to subsidence and severe problems arise in densely populated urban areas is Calatayud in Spain (e.g., Gutiérrez 1996). The main difference of the impact of such hazards compared to rural areas, and hence vulnerability, is the high abundance of infrastructure in urban areas as well as multiple anthropogenic activities.

Currently, protection concepts for groundwater and infrastructures basically have a monitoring character, while collected data correspond to historiography. However, integrated concepts have to consider local investigations in the context of regional urban landscape development and multiple interests with regard to surface and subsurface water use as well as protection that challenge the aims of water engineering and protection schemes. To provide a solid base for reliable risk assessment, adequate knowledge on different processes of evaporite karst evolution, its localization and its relationship to groundwater flow regimes and resulting land subsidence is required.

Generally, geology is understood as a steady-state and stable entity. However, due to the high solubilities of some rocks and the increase of hydrogeologic processes, geology can become transient and the role of water dominant. Whereas the occurrence of natural hazards within karst systems is not new for geologists, they often are surprised how suddenly land-subsidence and collapses can occur. This is mainly due to the relation of the high solubility of evaporite rocks in contact with water, especially when hydraulic boundary conditions change. Within this section, we concentrate on two case studies within densely urbanized river valleys.

We demonstrate that different investigative methods for karst process characterization in soluble rock complement each other and allow the interpretation of short-term impacts and long-term development on system dynamics. The obtained results show that models can be applied not only for theoretical research of simplified and idealized karst aquifers, but also to places with complex geological and hydrological properties. Investigative methods for similar subsidence problems can be optimized, leading from general measurements and monitoring technologies to tools with predictive character.

5.4.1 Karst Processes in Urban Areas

Generally, hydrogeological processes within karst systems are poorly understood. This has its reason mainly in the difficulty in collecting representative data within karst systems. On the one hand, karst systems are extremely heterogeneous and groundwater flow is complex. Groundwater flow in karst systems can be described by slow or diffuse and fast or conduit flow components. To obtain representative hydraulic data and to localize and capture the single components of the dual flow systems independently is very challenging. On the other hand, once representative data are available, modeling groundwater flow in karst environments poses an enormous challenge as hydraulic conductivities can span many orders of magnitude. As a result modeling results often are highly uncertain because of the lack of site-specific information on heterogeneous subsurface structures and the resulting complexity of flow paths. Structural features such as folds and faults can have further significant influence within karstified areas.

Karst processes are observed in the urban agglomerations east of Basel in the Folded and Tabular Jura as well as in the river valleys where gravel deposits are underlain by carbonate and evaporite formations. Under natural conditions karst features in carbonate rocks may develop over long time periods. However, under specific corrosive conditions and anthropogenic changed boundary conditions of groundwater systems karst features might develop much faster. Whereas rocks as limestones and dolomites are more soluble than many other rocks, the solubility of evaporites as gypsum or salt is even higher. With solubilities of up to 2 g l^{-1} for gypsum and even 358 g l^{-1} for halite (Kaltofen et al. 1994), these rocks belong to the most soluble rocks and evaporite-karst features develop much more rapidly than carbonate-karst features do.

5.4.2 Concepts and Investigation Methods

Theories that describe the evolution of karst systems are based mainly on conceptual models. For a more fundamental understanding of rock–groundwater interactions and the evolution of dissolution and flow within karst aquifers, modeling techniques have to be applied that are based on fundamental and well-established physical and chemical principles. There has been great progress in numerical modeling of the evolution of karst systems during the last decades. Today dissolution and early karstification of locations with complicated geological and geochemical settings can be modeled and the knowledge about basic processes governing dissolution and karst evolution has increased significantly. Different modeling methods are used for karst systems, whereas the various modeling techniques capture different aspects of the hydrologic processes. Table 5.1 summarizes the modeling approaches for karst environments.

Specific experiments help to extend the knowledge on some relevant processes taking place in karst systems and to construct the boundary conditions of modeling approaches. These experiments can be “active,” as hydrogeophysical surveys and

Table 5.1 The different modeling approaches for karst environments

Global	Distributive
Spatially nondifferentiated	Spatially differentiated
Analytical models that solve equations from hydrograph recession, breakthrough curves	Consideration of discrete faults (statistically or concrete)
Time series analysis	Consideration of double continuum
Numerical models (Black-Box)	Hybrid (discrete faults and continuum)
Little data needs	High data needs
Real structures are not considered	Consideration of real structures
Mathematically based	Mathematically and physically based
Reduced applicability for prognosis	Applicability for prognosis

laboratory experiments, or "passive," as observations and data obtained from natural or man-made hazards and infrastructure projects.

5.4.3 Case Studies

We present two case studies that allow the illustration of the complex interrelations between natural phenomena and processes that are induced by present day engineering and subsurface activities in the region of Basel. In the first case study, we illustrate the dissolution of deep-seated evaporites which leads to widespread subsidence in the investigated area. In the second case study, we highlight spatial and temporal aspects of karst evolution as well as water protection issues during remedial measures of a subsided highway and a river dam (Sect. 5.2), both constructed on gypsum-containing rocks. For the two examples we exemplify how a combination of monitoring, laboratory and hydrogeophysical experiments, as well as different modeling approaches can be applied to capture and understand a variety of relevant karst processes.

5.4.3.1 Land-Subsidence Caused by Intrastratal Evaporite Karstification

Dissolution of deep-seated evaporites leads to widespread subsidence in the densely urbanized area west of Basel (Fig. 5.19). As causes for the observed subsidence, several activities with possible impacts on hydraulic processes, starting at the beginning of industrialization, have to be considered:

1. Natural dissolution of the evaporites.
2. Salt solution mining.
3. Large-scale extraction of groundwater.
4. Connection of originally separated aquifers through drilling that potentially initiated evaporite dissolution processes as already observed in the region and discussed in the second case study.

The subsidence of a tunnel is only one example of a hazardous site of several known subsidence zones in the area. During an observation period of 78 days in 1997, subsidence occurred in a section of the open-mined tunnel at rates of 6 to 10 mm per month. Inclinometer and extensometer measurements showed that the vertical movements were located at the top of a salt horizon within the Middle Muschelkalk (Fig. 5.20). The Middle Muschelkalk includes the Upper Sulfatzone on top of the rock salt beds, which consists of strongly weathered, brecciated units and sequences of sandy marls and clays with intercalations of gypsum and anhydrite. The loss of drill-core material at the bottom of the Upper Sulfatzone is an indication for the existence of a salt karst zone at this level.

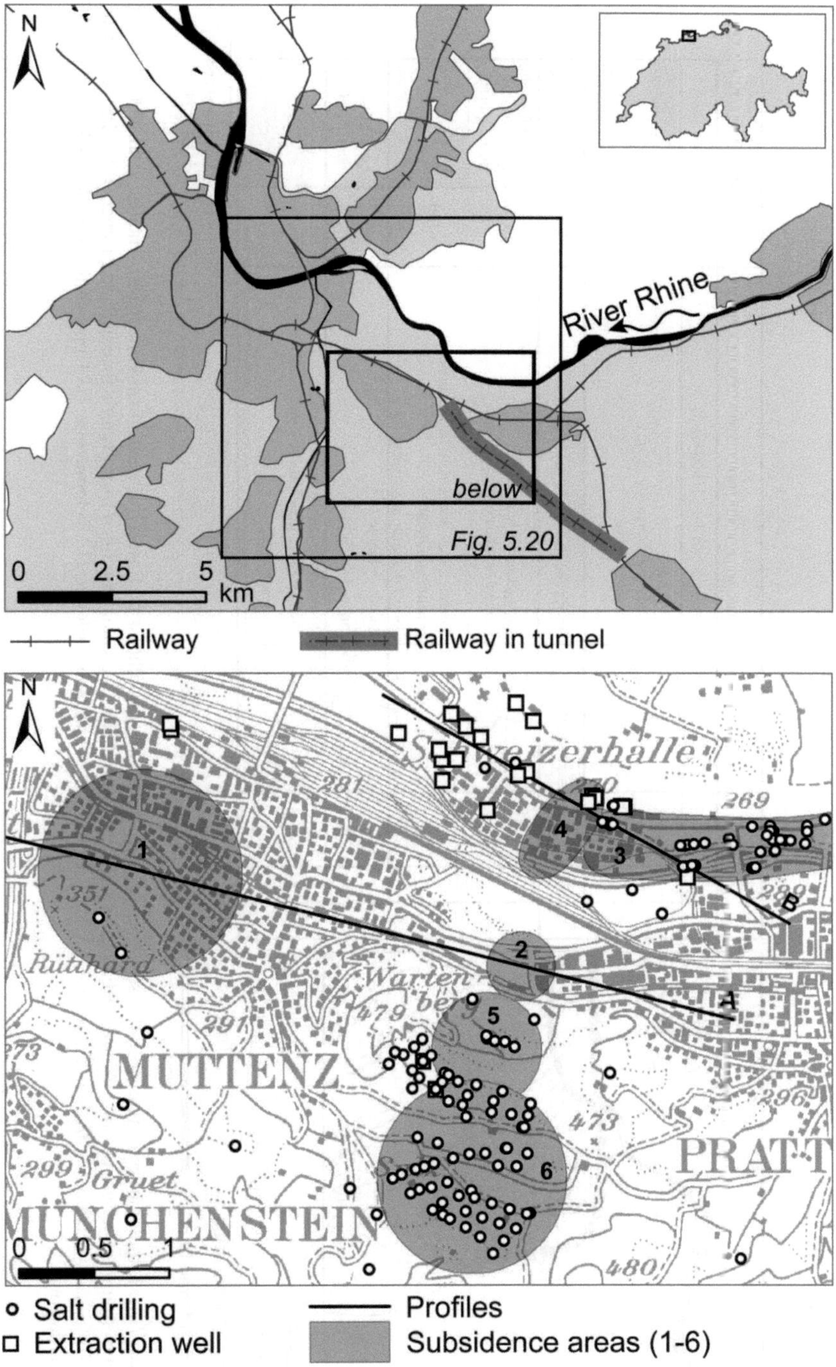

Fig. 5.19 *Above*: Location map showing the position of the new railway line *Below*: Land-subsidence areas (1 = Margelacker; 2 = Laachmatt; 3 = Ättigraben; 4 = Schweizerhalle; 5 = Hinterer Wartenberg; 6 = Zinggibrunn-Sulz), brine production fields and discharge wells in the Schweizerhalle region. The two hydrogeological cross-sections (A and B) are shown in Fig. 5.21. Topographic cards are reproduced with permission of swisstopo (BA110446)

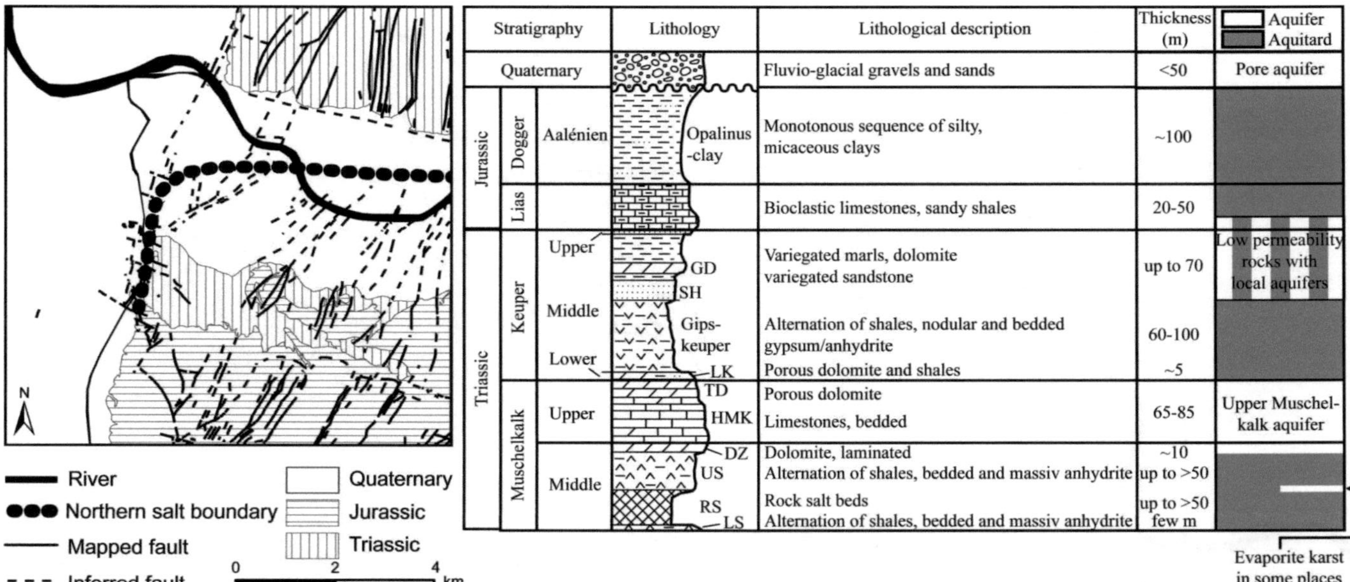

Fig. 5.20 *Left*: Lithologies and main tectonic structures of the Rhine valley southeast of Basel, Switzerland (modified from Hauber et al. 2000). *Right*: Schematic, stratigraphic section showing hydrogeological characteristics of the study area (modified from Bitterli-Brunner and Fischer, 1988; Pearson et al., 1991). *GD* Gansinger Dolomit, *SH* Schilfsandstein, *LK* Lettenkohle, *TD* Trigonodus-Dolomit, *DZ* Dolomitzone, *US* Upper Sulfatzone, *RS* rock salt layer, *LS* Lower Sulfatzone

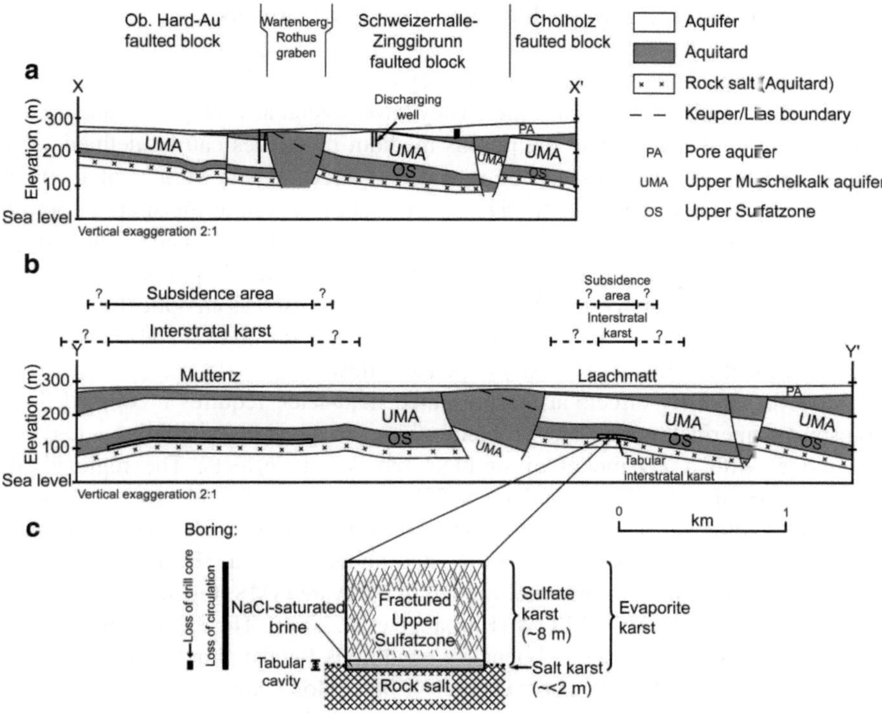

Fig. 5.21 Hydrogeological cross-sections A and B (lines of sections shown in Fig. 5.19). The blocks and graben are named after Herzog (1956). Some discharging wells are presented in A. In C a schematic close-up of intrastratal evaporite karst in Laachmatt is illustrated

Figure 5.21 presents a schematic, stratigraphic, and hydrogeological profile of the study area. To focus on the existing interstratal salt karst in the study area, the two cross-sections (oriented NW–SE and WNW–ESE) illustrate the existing geological and hydrogeological information including subsidence data. The line of section A is located close to the river Rhine, the line of section B cuts through both subsidence areas (c.f., Fig. 5.19).

Evaporite-dissolution occurs when groundwater, undersaturated with respect to the highly soluble minerals, enters into contact with an evaporite body. When groundwater interacts with a practically impervious, but highly soluble, evaporite body, the location of contact determines the resulting dissolution phenomena. The genesis and shape of the resulting deep-seated dissolution forms and the influence of fluid density on the transport and mixing processes are poorly understood. The illustrated experiments and modeling approaches that concentrate on groundwater interaction with salt formations from above include (1) a combination of flow tank experiments and numerical modeling to study the effect of variable density flow in heterogeneous media; and (2) laboratory dissolution experiments to study the effect induced by freshwater contacting salt.

Density Driven Groundwater Flow and Transport Models

Laboratory experiments are an excellent way of providing data to develop transport theories and validate numerical codes. They have several advantages: boundary and initial conditions are known, the porous medium properties can be defined separately, and the experiments can be repeated if necessary. A variety of methods exists for qualitative and quantitative determination of solute transport in porous media experiments. Konz et al. (2008) used dyes to qualitatively and quantitatively visualize spatial mixing patterns in an intermediate scale 2D porous media flow tank experiment. To study density effects the dye was used as an optical tracer. The benchmark experiments enabled to derive parameters for mixing and transport for the subsequent modeling of density dependent flow.

Modeling density effects at the presented field-scale requires a coupled flow-transport numerical model. The problem was approached by combining a regional 3D model with a 2D model to simulate the density effects. The regional 3D hydrogeological model is based on a 3D tectonic model and was constructed using the geological modeling software GOCAD (Spottke et al. 2005). The main aquifer-aquitard boundaries then were incorporated into a 3D hydrogeological model set up with the groundwater modeling software GMS, together with the 3D finite difference code MODFLOW (Harbaugh et al. 2000). The 3D hydrogeological model could be used to define boundary conditions for the simulation of density-coupled solute transport. An approximately 1,000 m long, and 150 m deep 2D field-scale model was used for a series of density-coupled solute transport simulations (Fig. 5.22).

For modeling the flow of fluids in fractured rock we chose a continuum approach where the fractured mass is assumed to be equivalent to a porous medium. We further assumed that preexisting fault zones at the WNW and ESE end of the cross-section, and the previously described interstratal salt karst unit were both 10 m thick. Hydraulic boundary conditions were assigned according to the regional 3D hydrogeological model: a prescribed constant head of 254.5 m a.s.l. was imposed on the ESE boundary. Anthropogenic groundwater withdrawal of a well field at the WNW end of the 2D cross section was simulated with a location corresponding to the pumping well with the largest withdrawal (up to 15 l s^{-1}). We simulated the pumping well with an evenly distributed constant flux along the corresponding vertical filter length and fitted to a piezometric head of 251.0 m a.s.l. as observed in the 3D simulations. Following the Hydrocoin test case (OECD 1988), a constant concentration boundary condition was applied with a density of 1,200 kg m^{-3} at the lower model boundary and assumed no-flow. Thus, salt can enter the domain only via diffusion what is considered to be a reliable approximation of the dilution kinetics of rock salt (halite). In case dissolution processes are purely transport controlled, ions are detached so rapidly from the surface of a crystal that they build up to form a saturated solution adjacent to the surface. Dissolution then is regulated by transport of these ions via advection and diffusion into the surrounding undersaturated solution. The chosen lower boundary condition at the top of the salt layer is based on the assumption that salt dissolution was instant compared to the

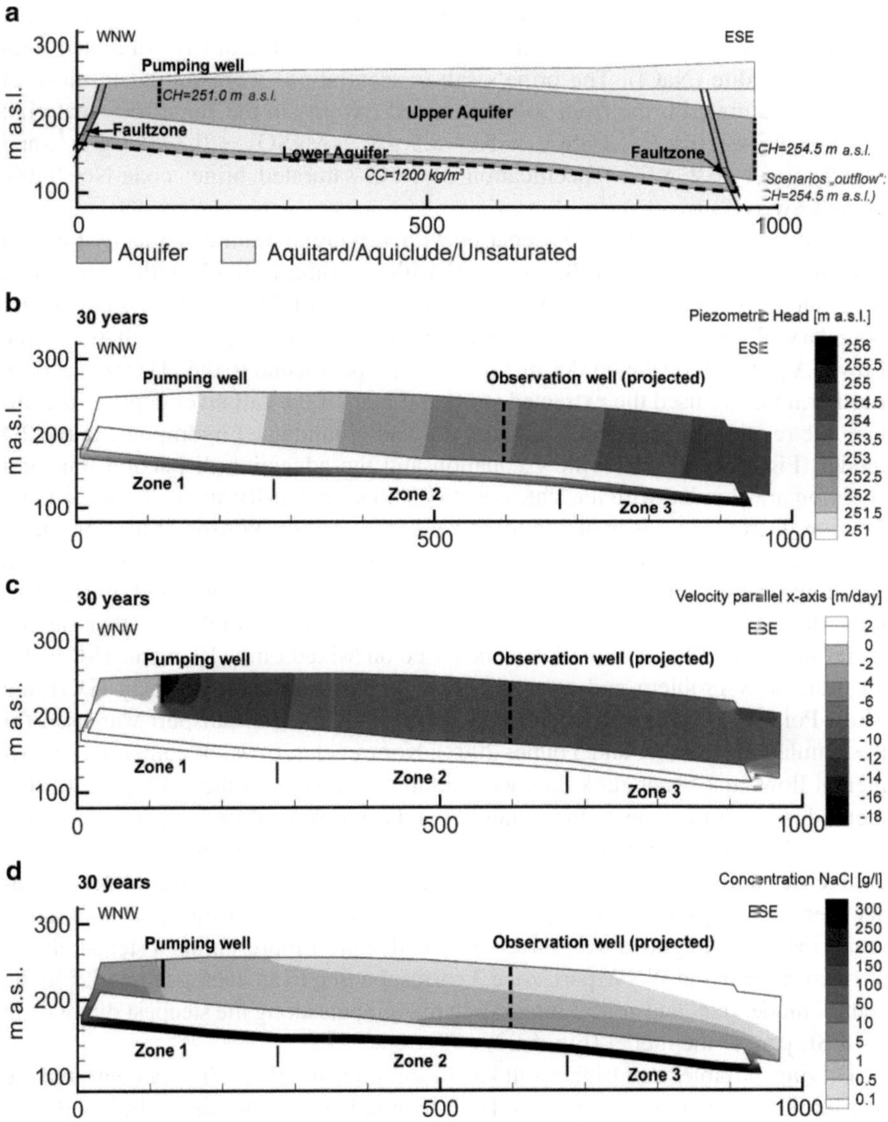

Fig. 5.22 (**a**) 2D cross-section with hydrostratigraphic model for the simulation of density-coupled flow and transport. Used boundary conditions are "Constant Head" (CH; *narrow dashed line*) to simulate pumping at the well, inflow from the ESE, and outflow at the bottom towards the ESE. Boundary conditions of "Constant Concentration" (CC; *wide dashed line*) simulate solute flux into the bottom of the Lower Aquifer. (**b**) Simulated piezometric head distribution from density-coupled model after 30 years, and location of zones 1–3 for integration of mass input. (**c**) Simulated x-component of velocity from 2D density-coupled model after 30 years. (**d**) Simulated concentration distribution from density-coupled model after 30 years

simulated time span of 30 years. Due to the fast halite dissolution kinetics, it can also be assumed that water in contact with rock salt is always saturated with respect to halite (NaCl). The brine's salt concentrations were similar to those of the NaCl-saturated brine from solution-mined caverns in the same rock material: $NaCl = 300–310\ g\ l^{-1}$, $CaSO_4 = 5.0–5.5\ g\ l^{-1}$, $MgSO_4 = 0.3–0.4\ g\ l^{-1}$, and $MgCl_2 = 0.1–0.2\ g\ l^{-1}$ (specification of NaCl-saturated brine; code-No. 9400; United Swiss Saltworks).

We conducted a total of 15 scenarios with varying hydraulic conductivities and porosities with three main differences (1) with pumping well, (2) without pumping well, and (3) with an outflow (fixed piezometric head of 254.5 m a.s.l) on the lower boundary of the model domain to simulate a possible drainage, or recharge of the Lower Aquifer along the bordering fault zone perpendicular to the 2D cross section. Furthermore, we used the extracted structural map of the salt layer deposit to compare the relationship between structural dip and groundwater hydraulics. In the 2D model (Fig. 5.22), Darcy's flow equation and the advection dispersion transport equation are coupled with the state equations, linking density and viscosity to mass fraction. Darcy's law is assumed to be valid as stated by Watson et al. (2002). At this stage, we assumed that the Fick's law is valid in order to avoid the use of an additional parameter. Because of this nonlinear coupling, codes based on standard numerical schemes require often very fine spatial and temporal discretization and therefore much CPU. A numerical code based on Mixed Finite Elements (MFE) for the fluid flow problem and a combination of Discontinuous Galerkin (DG) and Multi-Point Flux Approximation (MPFA) methods for the transport was used for the simulation (Ackerer and Younes 2008). Konz et al. (2009a, b) compared experimental flow tank benchmark data with simulation results of the MFE_DG_MPFA method in order to evaluate the reliability and robustness of the numerical scheme. The method turned out to be highly appropriate for simulations in complex geometries. Furthermore, simulated salt mass inputs along the lower model boundary were investigated. Mass inputs of all 15 density-coupled transport simulation scenarios were integrated over three zones with zone 1 representing a steeper dip of 5° on average in the WNW part, zone 2 corresponding to an average dip of 2° in the central model part, and zone 3 integrating mass inputs along the steepest dip of 6° in the ESE part of the model (Fig. 5.22).

Driving variables that trigger salt karst formation are hydraulic gradients and the distribution of hydraulic properties. High hydraulic conductivities paired with low porosities led to higher flow velocities which increase the concentration gradient in the lower aquifer and favor salt mass input along the bottom boundary condition. For the investigated setup, salt dissolution turned out to be enhanced by increased water fluxes along the formational surface of the salt layer. Considering the current boundary conditions, salt dissolution is boundary flux controlled. Steeper formation dips lead to an increase of fluid velocities and thus to an increase of fluid fluxes. Salt dissolution increases under the present boundary conditions of up to four magnitudes. This suggests that beside the hydraulic boundaries and the hydrogeological parameters, structural dip affecting density-driven transport plays an important role on providing the necessary hydraulic energy to support the process of salt karst formation.

The density-driven simulations along 2D cross-sections have limitations for a general interpretation regarding the causes for some of the observed land subsidence structures in the region of Basel. One limitation concerns the need for additional spatial and temporal data to describe the complexity of 3D structures as the top of the salt layer or the exact location or hydraulic properties of the fault zones as well as highly transient hydraulic boundary conditions. A second limitation derives from the assumption of a well-established 10 m thick homogenous evaporite karst layer which does not take into account the initial process of karst evolution. For instance, it is very likely that the groundwater extractions in the "Upper Muschelkalk Aquifer" accompanied by the creation of groundwater pathways in the salt bearing formations play a triggering role at the initial stage of salt karst evolution, whereas at a later stage structural setting dominate.

The recorded subsidence rates as observed in the tunnel indicate that phases of no or small vertical movement alternate with phases of accelerated subsidence. The transition between phases could represent the alternation of (1) steady-state conditions with little changes in the regional groundwater flow regime; followed by (2) salt dissolution phases as a result of changes within the regional scale hydrogeologic system and groundwater flow regime, due to (a) natural extreme hydrological events (floods within the catchment and breakthrough events within the karst system), or (b) anthropogenic activities, as salt mining, river dam construction and extensive groundwater extraction and injection within the region (Fig. 5.19, compare also to Sect. 5.3).

Salt Dissolution Experiments

To overcome the complexity of field measurements, we scaled down the study of fluid density effects at the top of a salt layer to laboratory dissolution experiments, where freshwater was slowly pumped from above into rock salt cores through an axial borehole (Fig. 5.23). Our experiments describe the effect induced by freshwater contacting from above the upper end of the core cylinder (top of the salt layer) showing a deviation from the perfectly horizontal layered rock. The Plexiglas/silicone top cover functions as a practically impermeable and poorly soluble stratum (covering layer). A schematic diagram of the experimental setup is illustrated in Gechter et al. (2008). Compared to a field situation, the main laboratory constraints were (1) the supply of NaCl-undersaturated water through a pipe (quasi-1D input) and not through a more realistic linear input (permeable fault) and (2) the dissolved salt water was transported through the salt core (salt layer) instead of being transported upwards/laterally away from the top of the salt layer such as in an overlying freshwater aquifer.

Since faults displace salt horizons in horst and graben structures, and may constitute very important types of discontinuities in rocks from a hydrogeological point of view, further experiments evaluated the influence under more realistic

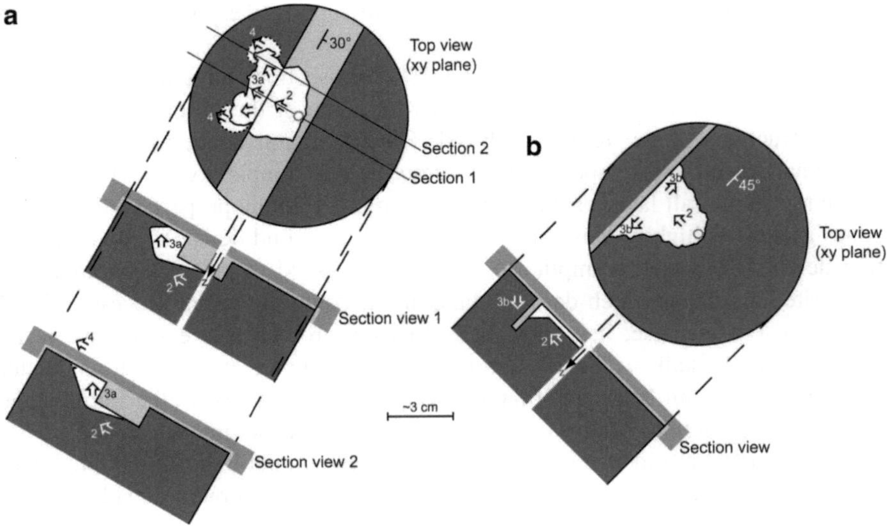

Fig. 5.23 Schematic diagrams: Solution cavity for (**a**) first type and (**b**) second type of experiments

geological conditions. We studied two cases (there are endless variations): a normal fault juxtaposing a developing interstratal salt karst against a salt layer, and a normal fault juxtaposing a developing interstratal salt karst against a largely impermeable, poorly soluble layer (Fig. 5.24). After unmounting the drill core from the holder system, we examined the undisturbed final solution cavities macroscopically.

In the first type of experiments, the average inflow rate amounted to ~10 or ~80 mm^3 min^{-1} over time scales of 3.4–31.2 h, and the inclination of the top of the salt layer varied between $<1°$ and 45°. In the second type of experiments, the aforementioned Plexiglas/silicone top was extended by a silicon-filled groove placed parallel to the strike of the inclined top of the salt layer in order to simulate a horst and graben structure (displaced layers; fault). In both dissolution experiments, the upper top of the salt layer was relatively steep parallel to the layers (30 or 45°, respectively) and the average inflow rate was ~70 mm^3 min^{-1} over 5.2 or 11.6 h, respectively.

The experimental solution cavities show a close correlation to cavity geometries observed in the field. Dijk and Berkowitz (2000) described the development of cavity geometries as preferential dissolution that occurs at higher elevation of a developing cavity leading to a cavity expanding selectively into the overlying salt rock and a broadening of the cavity upwards. Accordingly, the experimental solution cavities indicate a development of the interstratal salt karst along the top of the rock salt layer at the beginning, mainly in the up-dip direction by a sideward displacement of the main dissolution front. When in contact with a fault, an areal horizontal salt dissolution front mainly broadens the solution cavity downwards or upwards, leading to a fault-associated solution cavity (section) possibly responsible for the overlying fault-parallel subsidence bowl. The results of the dissolution

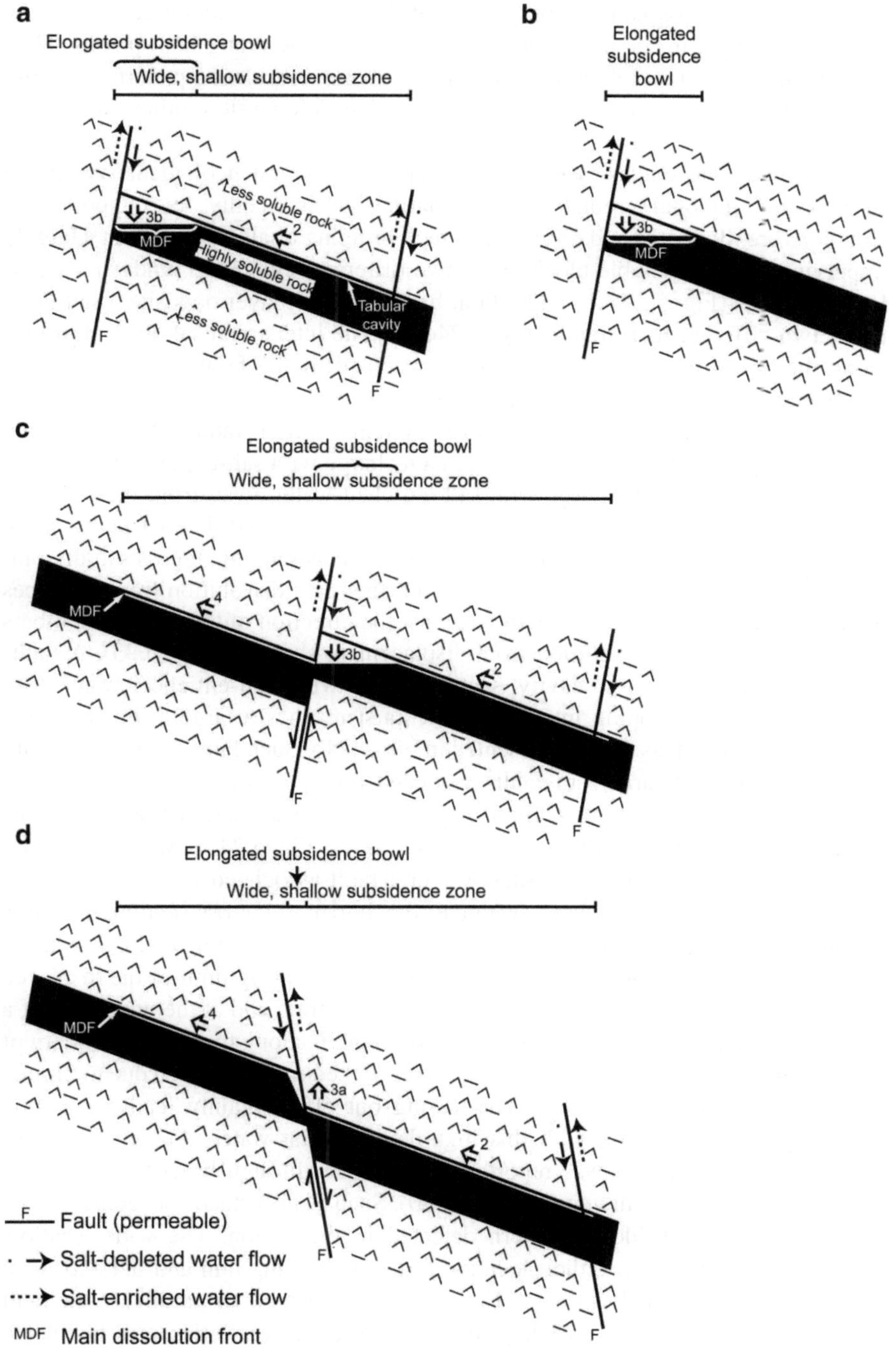

Fig. 5.24 Possible developments (**a**–**d**) of an intrastratal salt karst at the top of a salt layer in the context of extensional tectonics and inclined beds

experiments allow explaining the genesis of flat expanded solution cavities at the top of a salt layer inside horst and graben structures and along fault planes. They may be also used to explain some of the overlying subsidence patterns of those areas which are not directly related to the dissolution salt mining subsidence (Fig. 5.19).

Since there are endless variations to represent natural geological subsurface settings and regional groundwater circulation systems, only four hypothetical situations are illustrated and discussed: a fault bringing a salt layer against a largely impervious, poorly soluble block under two different regional groundwater circulation systems (Fig. 5.24a, b) and a fault bringing a salt layer opposite a salt layer under two different fault trends (Fig. 5.24c, d). The interstratal salt karst develops along the top of the salt layer analog as presented by the flow experiments. The permeable normal faults and inclination of the top of the salt layer clearly influence the development and spread of this karst system. Normal faults act as preferred channels of groundwater flow resulting in a recharge with salt-depleted water from above to the salt and discharge with salt-enriched water away from the salt. The movement of these two solutions occurs along separate and distinct paths in the same fault(s) and/or along respective fault(s). At the beginning, unless salt dissolution starts at the up-dip limit of the salt layer, the main dissolution front displaces along the top of the salt layer, mainly against the direction of the dip (2) (numbers refer to the direction of solution front displacement shown in Fig. 5.23). A fault-associated, large solution cavity section filled with salt-enriched water (3b) develops if the developing interstratal salt karst reaches a largely impermeable, poorly soluble block as the one presented in Fig. 5.24b. Such fault-associated cavity geometries are comparable to fault-associated oil traps found in the petroleum geology. Such a developing section may also lead to a continuation (4) of this karst system along the top of the displaced salt layer (Fig. 5.24c). Another continuation (3a) is presented in Fig. 5.24d; however, a fault-associated salt-saturated filled solution cavity developed here, whereas the horizontal cross-section increases upwards. Salt dissolution occurs mainly at the sideways displacing main dissolution front (2, 4) or horizontal main dissolution front at the up-dip limit of the displaced salt layer (3b) or simultaneously at the horizontal ceiling and at the upper end of a facet (3a). Aside from the main dissolution front of 3b, along the lower boundary of the tabular cavities only very small dissolutional receding is observed due to limited advective and diffusive mixing. The salting out effect mainly leads to carbon dioxide gas in association with this high-density karst water. At field scale for example, such degassing was reported from drilled through brine-filled solution cavities at the top of a salt unit (Belchic 1960). Such a subsurface conceptual model may explain the subsidence pattern detected in the region. The wide, shallow subsidence zones may owe their location to the underlying thin and flat expanded section of the interstratal salt karst. Elongate subsidence bowls parallel to the fault system within these zones, revealing stronger subsidence velocities, may be related to underlying, fault-bound (higher) solution cavities. A fault may also act as a subsidence barrier as shown in Fig. 5.24a, b.

On the geological maps only major faults, which are observable on the surface, are mapped. According to the results of the dissolution experiments the maximum subsidence of the elongated subsidence zones parallel to the strike of the major faults could be an indicator for smaller scale fault systems as have been observed in tunnel excavations. Hence we postulate that further fault systems exist next to the main mappable tectonic structures and that the observed dissolution mechanism is more frequent. Our findings of the flow experiments can be upscaled and integrated into a conceptual model for interstratal salt karst development by density-dependent groundwater flow and corresponding overlying land-subsidence patterns in the context of horst and graben structures as for example observed in the region.

The investigations we presented reveal the advantage of small-scale dissolution experiments and numerical modeling of density driven groundwater flow when trying to understand the complex processes underlying natural (and possibly anthropogenic induced) intrastratal salt karstification. Our experimental data allow explaining (1) the genesis of flat expanded solution cavities at the top of a salt layer inside horst and graben structures; and (2) to some degree the overlying subsidence patterns.

In the investigation area, different subsidence phenomena can be observed which still require a solid scientific base for explanation that is necessary to formulate concrete steps for an action plan. Until very recently a fragmented geodetic monitoring program did not allow a regional scale comparison and interpretation of the subsidence rates and patterns. In order to predict further subsidence and to plan appropriate measures, it was necessary to understand the current stage of the hydrogeologic flow regime as well as the subsidence mechanism and its development over time.

An appropriate morphological land surface analysis of an affected area (time series of subsidence, spatial distribution) would allow obtaining an overview of the distribution and the development of the underlying interstratal salt karst systems as well as the development of the roof of solution caves. Therefore widespread leveling measurements of land-subsidence are required if morphologies are difficult to detect visually, such as in the region.

The results of this research we will use as a basis for further experimental and conceptual model developments on salt dissolution and solute transport by density-driven groundwater flow. These may ultimately help provide predictions on land subsidence patterns and related risks of surface deformation. Whereas in future theoretical aspects of density-dependent groundwater circulation in contact with an intrastratal salt karst and its influence on rock mechanical behavior (subsidence of the overlying geological materials) has to be considered as a coupled process.

5.4.3.2 Simulation of Karst Evolution

The particular geological situations encountered during urban infrastructure development often are not considered in the planning phase. Such situations occur when infrastructure is constructed on soluble geologic formations that are prone to subsidence. The situations are even aggravated when groundwater flow regimes are altered and solution processes enhanced. When located close to artificial hydraulic structures such as river dams, natural processes of karstification can ensue due to (1) the presence of evaporites; (2) elevated hydraulic gradients; and (3) the presence of undersaturated water. Such boundary conditions can lead to increased leakage, to subsidence and to the failure and collapse of the hydraulic structures themselves and of nearby infrastructures.

In this case study, we present a possible implementation of different quality field data into hydrogeological karst models. We collected data in the context of an upgrading project for a highway located beside a river dam that was constructed on gypsum-containing rock (for details on the engineering part of the project and the settings please refer to Sect. 5.2B). The primary project goal was to develop tools that enable a continuous characterization of the groundwater flow regime and that facilitate the evaluation of the long-term performance of the highway and the river dam.

We present an approach which merges high-resolution 3D hydrogeologic modeling with 2D karst evolution modeling (Fig. 5.25). The different modeling techniques capture different aspects of the hydrologic processes. The hydrogeological model was presented in Sect. 5.2B and includes a deterministic finite difference approach which takes into account an equivalent porous medium for weathered and nonweathered rock, and a coupling of the system with drains that represent a generalization of the conduit component of model outflow (mixed flow in karst settings).

In the present case study, we approached the simulation of the evolution of flow within the gypsum karst aquifer by a simplified 2D finite difference method where hydrodynamic flow is coupled with equations of dissolutional widening.

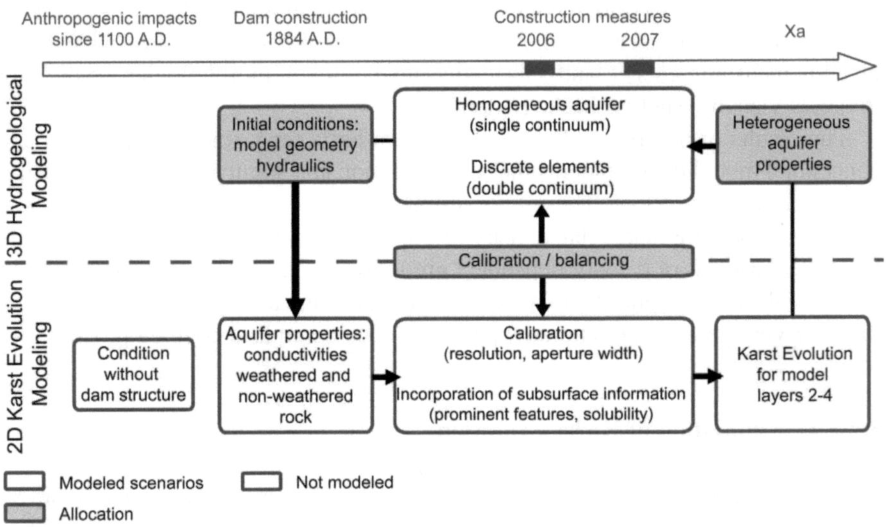

Fig. 5.25 Conceptual approach

The evolution of 2D karst aquifers has been intensively studied during the past decades, also close to dam structures (e.g., Romanov et al. 2007). These previous approaches to modeling karst evolution focused on hypothetical karst catchments with synthetic data. This study is one of the few studies that have attempted to integrate field data, which are sensitive to groundwater hydraulics, into real world applications which are related to locally specific issues.

Our main goals in setting up the 2D karst evolution model were:

1. To simulate the spatiotemporal development of karst features within the investigation area over the last 100 years.
2. To determine and evaluate the relationship of the investigated parameters.
3. To determine time scales for future system development.
4. To provide heterogeneously distributed aquifer properties including distinct high conductive features for the 3D hydrogeological model.

The reason for not modeling karst evolution with existing 3D codes is the complexity of model geometries and chosen boundary conditions that for the moment cannot be used without unreasonable effort and CPU. However, as 3D codes are gaining in efficiency, the available data sets will be transferred to 3D karst evolution model in the near future.

The 2D karst evolution model includes representative geological horizontal cross sections of the area around the dam structure. With respect to the high information density in the vicinity of the infrastructures, horizontal cross sections were preferred over the generally used vertical model setups. The geometric dimensions as well as the hydraulic and technical boundary conditions (modeling domain, groundwater and river head, hydraulic conductivities) of the 2D karst evolution model correspond to those of the 3D hydrogeological model (see Sect. 5.2B). The modeling domain is 130 m wide (W–E) and 465 m long (S–N). Figure 5.26 shows the conceptual model setup, the impervious and insoluble dam structure and zones used to assign different hydraulic conductivities and solubilities. Flow within the karst aquifer is driven by infiltrating river water upstream of the dam that is directed towards the river downstream of the dam as a base level. In advanced scenarios, regional groundwater flows from South to North and from the adjacent slope to the East were considered also.

The application of gypsum dissolution kinetics (Dreybrodt 1990) enables the simulation of void and conduit enlargement with time by chemical dissolution and solutional widening within the system. This selective enlargement increases conductivity by several orders of magnitude during the early phase of karstification. Thus, flow characteristics change from more homogeneous to heterogeneous flow. An important moment for the stage of karst aquifer evolution is the so-called breakthrough time, when flow changes from laminar to turbulent and increases by several orders of magnitude in a relatively short interval.

In the present case study, evolution of distinct karst features should be in the range of 10–100 years, because:

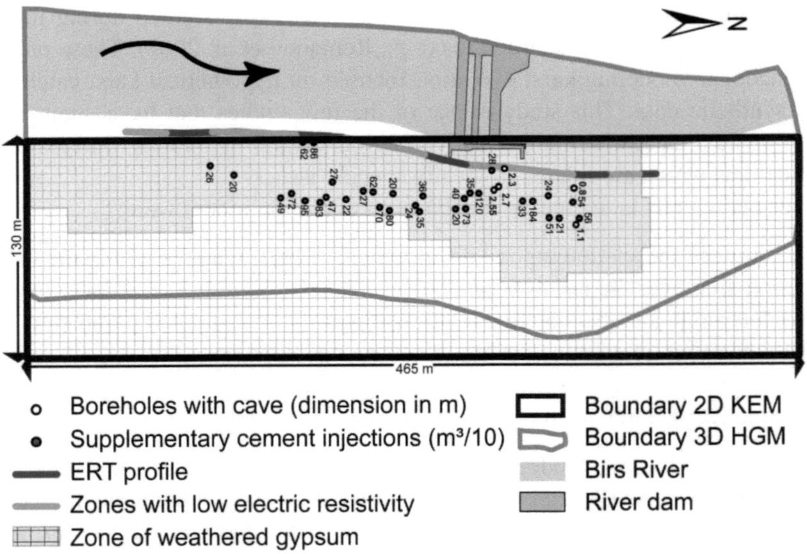

Fig. 5.26 Geometry and hydraulic boundary conditions of the horizontal 2D karst evolution model superimposed on the model geometries of the 3D hydrogeologic model. Illustrated also are zones for weathered and nonweathered aquifer properties as well as locations with information from electrical resistivity tomography, caves and injections of supplementary cement

1. The dam structure in its current dimension was built in the 1890s.
2. Subsidence of the highway has been reported since the 1990s.
3. Previous dam site modeling indicates that under normal conditions, evolution periods of several thousands of years can be shortened to periods of several decades for karst aquifers in the vicinity of hydraulic structures such as dams.

As structural initial conditions can influence karst evolution in a significant way, subsequent scenarios focused on the incorporation of specific subsurface information for the description of aquifer heterogeneity. These scenarios include statistical distributions of hydraulic conductivity and solubility zones and discrete prominent fractures.

We discuss three time intervals of the aquifer's development in detail. The first covers the natural karstification for a period between several hundred up to a few thousand years. The results from this evolution period are used as initial conditions for the second interval, which covers the time between 1890 and 2007 AD. This period is characterized by anthropogenic alterations of the system through the man-made river dam, which considerably changes the evolution of the aquifer. In 2006 and 2007 AD – after serious subsidence of a nearby highway has been observed – technical measures have been conducted and thus the boundary conditions have changed once again. This is the beginning for the third modeled interval. A forecast for the following 100 years is developed.

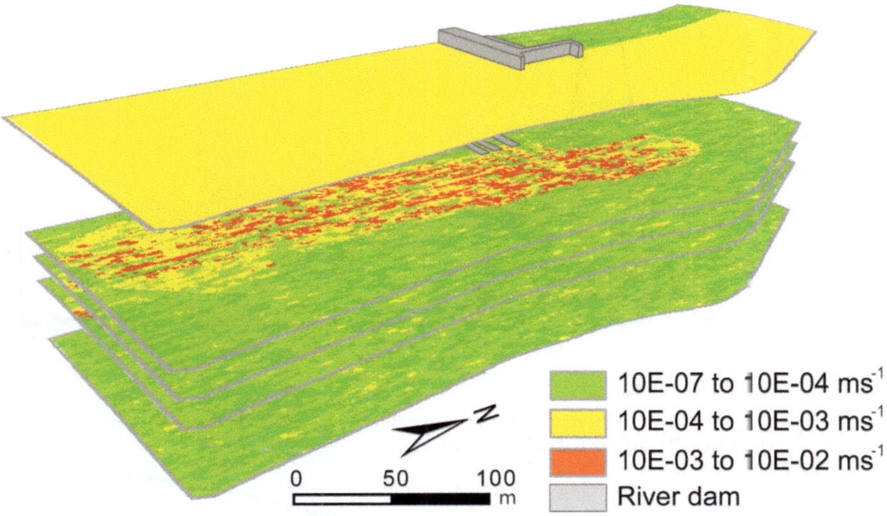

Fig. 5.27 Visualization of heterogeneous hydraulic conductivity distributions transferred from three realizations of the 2D karst evolution model to 3D hydrogeological model (Layers 2–4). Aquifer properties distributions for Layers 1 and 5 are described in the text

In the following, we discuss the results of one scenario, including statistical distributions of hydraulic conductivity and solubility zones, based on information on locations where caves were encountered and supplementary cement was injected during the restoration of the highway, as well as where electrical resistivity tomography resulted in low resistivity, and of regional hydraulic gradients.

Figure 5.27 shows the statistical distributions of hydraulic conductivities for the nonweathered and weathered rock used for the setup. The development of aquifer properties from the nonweathered to the weathered state for the period of natural karstification can be derived from the shift of the statistical distributions for hydraulic conductivities and solubilities. For the generation of statistical distributed properties of the aquifer (aperture width and solubility), spatial correlation lengths were incorporated which characterize the geometric anisotropy of the sedimentary structure types. Spatial correlation lengths of 1/10 of the longitudinal (i.e., 46.5 m S–N) and lateral (i.e., 13 m W–E) model domain were chosen. Values for the solubility of the nonweathered and weathered zones were averaged from borehole descriptions, resulting in 15% of soluble fractures for the weathered and 40% of soluble fractures for the nonweathered zone.

Figure 5.28 illustrates the development of hydraulic conductivities from the initial state at 0 years (i.e., the 1890s) to 100 years (i.e., the 1990s). Please note the logarithmic scale of conductivities on the *y*-axis. While the gross distribution of conductivities does not change significantly, the resulting curve after 100 years evolution time illustrates (1) a slight shift from lower to higher conductivity

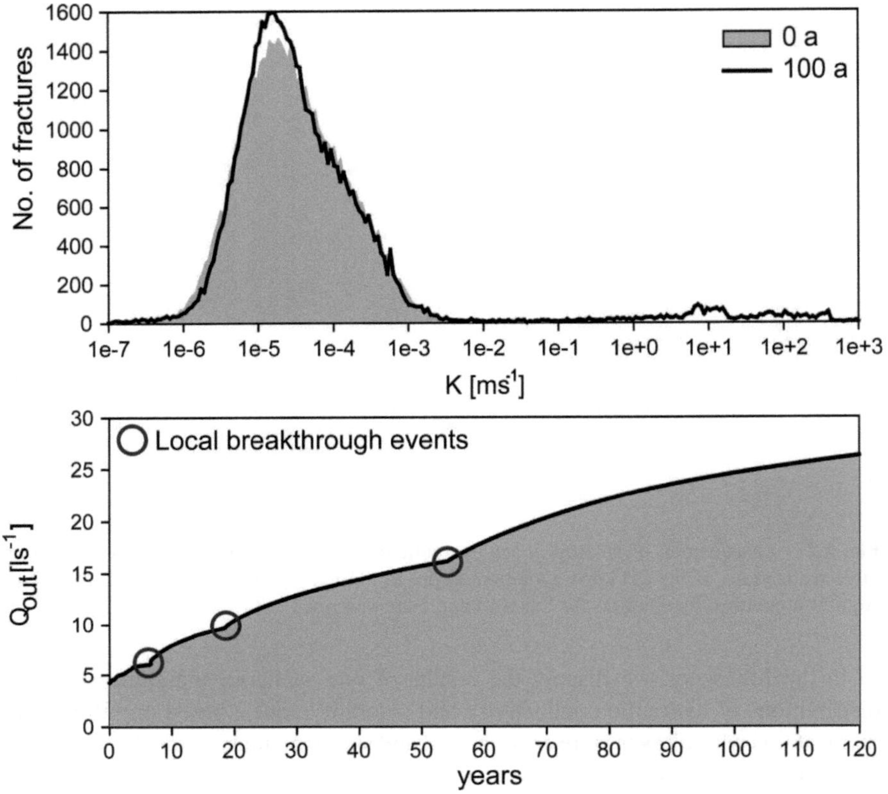

Fig. 5.28 *Upper graph*: Simulation results of karst evolution illustrated by the change of hydraulic conductivity distribution from the initial state to 100 a. *Lower graph*: Simulated model outflow

values; and (2) the development of a few fractures with very high conductivities. Comparing this to the results from simulated model outflows, it is obvious that the occurrence of a limited number of highly permeable structures and their interconnection can dominate the flow processes.

Figure 5.28 further illustrates the development of leakage around the dam for the 100-year time period from the 1890s to the 1990s. The graph shows a stepwise progression, while modeled outflow increases more rapidly at the beginning of the modeled time period, as remaining soluble zones within the weathered rock are dissolved. Subsequently, time spans between single steps increase. The single steps can be interpreted as local breakthrough events that lead to an abrupt increase of outflow. A steady increase in outflow can be observed between the single steps. As for the 2D karst evolution model average boundary conditions are chosen, modeled outflow after 100 years evolution time, resulting in approximately

$24 \, l \, s^{-1}$, can be compared to the lower values of modeled outflow resulting from the 3D hydrogeological model. Outflow simulated with the transient 3D hydrogeological model resulted in values between 6 and $13 \, l \, s^{-1}$ for the general head boundary downstream and between 12 and $77 \, l \, s^{-1}$ for the drains (cf. Fig. 5.14). Hence, modeled outflow with both modeling approaches during low to medium hydrologic boundary conditions are in the same order of magnitude.

The results from the 2D karst evolution model clearly illustrate that the amount of gypsum within the nonweathered and weathered rock can inhibit karstification and the development of connected percolation pathways necessary for breakthrough from infiltration locations to base levels. Moreover, local breakthrough events lead to delimited subsidence events as has been observed within the real world. The outflow progression, resulting from the 2D karst evolution model and heterogeneously distributed solubility, can illustrate patterns of karstification within gypsum rocks.

Our results correlate very well with the findings of the field studies of the area. Furthermore, predicted evolution timescales are reasonable from what is known about the past of the aquifer. The karst evolution models allowed us to simulate the development of aquifer properties, which subsequently could be transferred to the 3D hydrogeological model, allowing a more realistic representation of subsurface heterogeneities.

Long-term evolution of karst systems and prognosis could only be achieved by setting up models that account for the change in hydraulic properties with time. Our results presented in this case study show how calibrated 2D karst evolution model can be used to describe the current state of karst systems, and for prognosis of system development and subsidence risk assessment in the near future. The calibrated 2D karst evolution model allowed the karst aquifer evolution to be simulated to its current state and the sensitivity of changing natural and anthropogenic boundary conditions to be investigated.

5.4.4 Conclusions

The development of instruments that can be used for prediction requires the design of surface and groundwater monitoring systems, goal-oriented field campaigns, and selected modeling strategies in order to investigate the locally specific relevant processes. The strength of such models lies in the assessment of evolution time scales and the quantification of rates at which processes operate.

In addition, the development of infrastructures in areas prone to subsidence requires special sets of rules and regulations (e.g., prohibiting the connection of aquifers) to minimize potential problems from present and future development. Our described models require a step-wise approach, can have a predictive character and can be the basis for the development of effective long-term strategies within transient hydrogeological environments.

5.5 Geothermal Energy

Christoph Butscher, Jannis Epting, and Peter Huggenberger

Over the last decades, the thermal use of the urban subsurface and groundwater resources has increased significantly. Especially the use of groundwater for cooling and heating purposes has led to considerable pressures on urban aquifers. Moreover, the extension of city infrastructures into depth and diffuse heat input of heated buildings have resulted in elevated groundwater temperatures in many urban areas (see Sect. 3.4).

The number of geothermal energy systems is increasing rapidly in urban areas, mainly in residential areas (Fig. 5.29). Even though the environmental benefit of such systems with respect to reducing CO_2 emissions is unquestioned (Blum et al. 2010), there are environmental risks related to the installation of geothermal

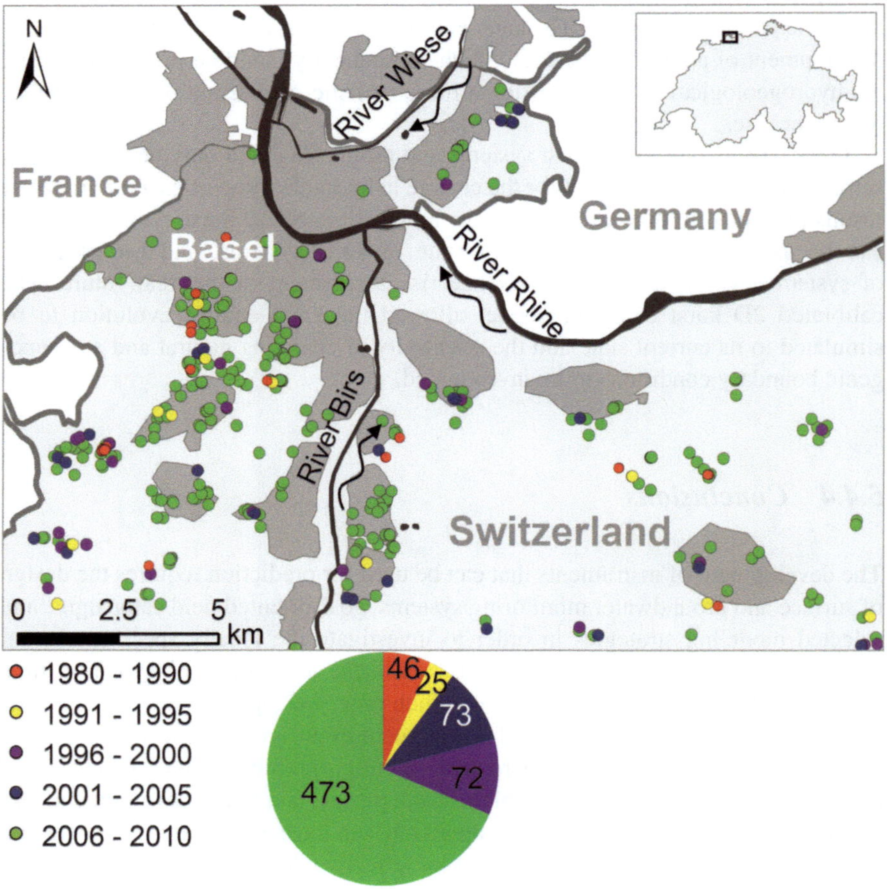

Fig. 5.29 Development of geothermal systems in the region of Basel

systems. Up to now no knowledge on the long-term interferences of geothermal use with the subsurface environment exists. Therefore, in the future, a higher density of geothermal use will lead to unavoidable conflicts between neighboring sites and other utilizations of subsurface resources.

Especially in complex geological and urban settings the continuous increase of shallow geothermal system installations bears some danger. For example, both the transformation of anhydrite into gypsum, which is accompanied by a considerable rock volume increase, and the leaching of salt bearing formations, can lead to terrain uplift and, respectively, land-subsidence. These are two major risks related to shallow geothermal technologies. Changes in rock–groundwater interactions induced by engineering activities that lead to negative impacts can only be prevented when the geological and hydrogeological settings of critical areas are known and when information is available in early planning phases.

Increasing geothermal groundwater use can exceed the subsurface potential for different heating and cooling systems and affect groundwater quality. Currently, in most urban areas, regulations for water resource management and geothermal energy use are sparse and often limited to the rule "first come, first served." As a consequence, groundwater temperature can increase significantly, as has been observed in northwestern Basel, where groundwater temperatures reach seasonally up to 17°C (approx. 10°C long-term average annual air temperature; see Sect. 3.4). While consequences of climate change are difficult to detect in groundwater resources and remediation requires global approaches, regional and local pressures are often orders of magnitude larger, in particular in urban areas. Negative effects of geothermal energy use can, however, be minimized by water resources management, provided that there are sufficient data for planning and risk evaluation. Such data is of particular importance in areas with evaporite formations, which are especially susceptible for rock swelling and subrosion.

In view of the risks associated with geothermal installations, guidelines are necessary for the administrative proceeding regulating the construction and operation of such installations (Butscher et al. 2010). In order to be able to trade off the expected benefits against possible risks, knowledge about the local geological characteristics, hydrogeological conditions, and the processes taking place is required. Geological and hydrogeological information can be provided to authorities and the public by suitability maps for geothermal subsurface use. The investigation of the processes downstream of thermal groundwater use and infiltrating surface waters can be accomplished using numerical heat-transport models.

In this section, we outline two of the most important aspects of geothermal energy use in urban areas, which include:

1. Consequences of connecting previously separated aquifers and rock–water interaction in evaporite formations after the installation of shallow geothermal systems.
2. Degradation of groundwater quality due to solution processes and heating up of aquifers.

In the first case study, we present concepts that allow rapidly evaluating the suitability of proposed drilling sites for geothermal use. In the second case study, we present a thermal groundwater management concept for the northwestern area of the city of Basel, which will serve as a decision tool for the sustainable thermal use of groundwater in urban areas.

5.5.1 Geothermal Settings and Boundaries

Generally, a distinction is made between deep (>~400 m) and shallow (<~400 m) geothermal energy use (Fig. 5.30). The use of deep geothermal energy includes engineered geothermal systems such as "hot-dry-rock" and "hot-wet-rock" systems (Nakatsuka 1999) as well as the use of water from deep hot aquifers. Deep geothermal energy projects are large-scale projects that are relatively rare in urban areas, and not subject of this section. However, in the Basel area it was planned to use deep geothermal energy. The outcome of this project is summarized in Sect. 5.6. In contrast, shallow geothermal systems are widespread in urban areas, and in view of increasing demand for sustainable energy it is expected that the use of shallow geothermal energy in urban areas will become even more frequent in the future (Sanner et al. 2003).

The use of shallow geothermal energy can be classified into closed and open systems. In closed systems, a heat transfer liquid (e.g., glycol–water mixture) is circulating in a closed loop, and the heat energy of the transfer liquid is gained using a heat pump. The most common closed systems are borehole heat exchangers (BHE), where a double u-shaped loop containing a heat transfer liquid is installed in a

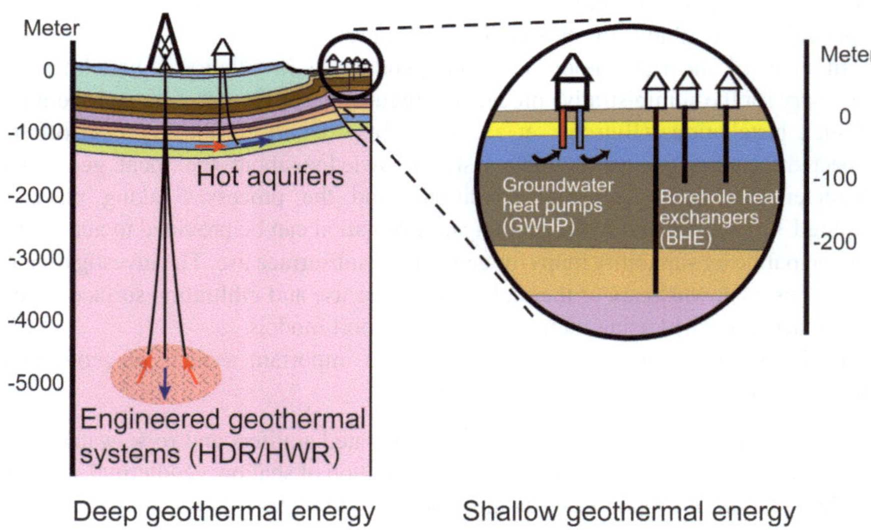

Fig. 5.30 Categories of different thermal systems: Deep (*left*) and shallow geothermal energy (*right*). Within shallow geothermal systems, a distinction is made between open (ground water heat pumps) and closed systems (e.g., borehole heat exchangers)

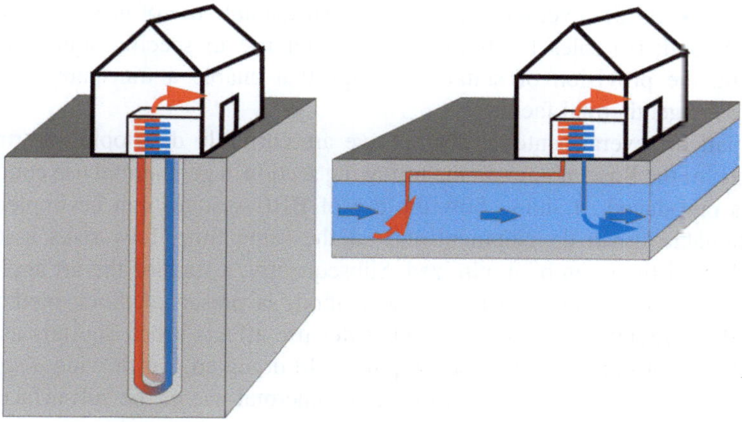

Fig. 5.31 *Left*: Closed thermal system. *Right*: Open thermal system

vertical borehole approx. 50 to 400 m deep (Fig. 5.31). Such systems are increasingly used for heating and cooling of buildings, which typically include small units (1–2 family houses). But also larger units (e.g., public buildings) use BHE systems, implying fields with several BHE. Open geothermal systems use the geothermal energy of the groundwater by means of groundwater heat pumps (GWHP). In contrast to closed systems, the groundwater itself is the heat transfer liquid. Such systems involve a pumping well to withdraw groundwater and a reinjection well where the used groundwater is reinjected into the aquifer after its thermal energy has been gained using a heat pump (Fig. 5.31). GWHP systems are mainly operated by larger (>50 kW) industrial energy users for heating and, more commonly, cooling purposes. The direct use of groundwater represents a large-scale interference with the aquifer, both with respect to the aquifer's thermal and flow regime.

Currently, for the cantons Basel-Stadt and Basel-Landschaft the geological factors and thermal groundwater regimes are reviewed. These factors play an important role in the assessment of the geothermal resource potential. The number of geothermal systems in the region of Basel increases steadily. Currently, approximately 700 sites for geothermal energy use exist in the region of Basel. Most sites (473) were installed after 2006 (Fig. 5.29).

During construction and the following operation of geothermal systems, the environment and especially groundwater resources can be endangered. Especially open systems for thermal groundwater use can have major influence on local and regional thermal groundwater regimes. This situation aggravates in urban areas where extensive groundwater use and the radiation from building parts reaching into the groundwater cumulate. Considering the sum of all these hydrological, hydrogeological, and operational boundary conditions the situation in urban areas is extremely complex.

Therefore, groundwater resource and thermal energy management concepts require knowledge on the relation between groundwater flow and temperature regimes, as well as on mixing under variable conditions. Such concepts include the definition of (a) development goals for a sustainable use of urban groundwater resources and; (b) rules for future groundwater use in specific aquifer regions, including the provision of suitability maps that enable a transparent approval practice for geothermal facilities.

As part of several cantonal projects we are currently developing thermal use concepts for the Basel region. In the following section, a geothermal use concept for BHE is introduced. It shows how the use of BHE systems can be implemented in sustainable urban development and, at the same time, how risks associated with BHE drillings can be minimized. Subsequently, a state-of-the-art application of advanced monitoring and modeling methods is presented. Such methods are required to predict how thermal groundwater use affects urban aquifers as a pre-requisite for sustainable urban development. In doing so, we provide a scientific basis to address several questions for the geothermal use of the subsurface in the urban area of Basel:

1. Which are the determining hydrological and hydrogeological processes, including their spatiotemporal scales that have to be considered near shallow geothermal use?
2. What thermal, chemical, and microbiological effects occur downstream of thermal groundwater use and how can they influence future groundwater use?
3. To what degree can water supply and thermal groundwater use systems be optimized?
4. How can a series of local water supply and thermal groundwater use systems be integrated into a network based on local and regional scale risk minimization, considering long- and medium-term development (development of groundwater and heat use concepts, suitability maps)?

5.5.2 Implementation of Geothermal Use Concepts for Borehole Heat Exchangers

In this section, we present a risk-oriented geothermal use concept which can be used for regulation and the evaluation of new geothermal systems. The concept considers potential threats related to the installation of BHE systems. The basic principle of the approach is that in regions with complex geological situations a more restrictive approval practice is recommended. In regions where comprehensive geological data are available, more differentiated solutions may be applied. According to this regulation, three different cases are possible: At a site, BHE systems are (A) not allowed, (B) allowed with specific requirements, or (C) allowed with standard requirements. Which of these options is the case at an actual site is defined by criteria meeting the concerns that are related to the installation and operation of BHE systems. The criteria are summarized in Table 5.2.

The criteria are related to the specific geological characteristics of a region or to regional land use planning within an urban environment. We present different

Table 5.2 Decision criteria for the approval of borehole heat exchange (BHE) installations applied by the public authorities in Canton Basel-Landschaft

Criteria	BHE installations are
Groundwater protection zone Contaminated site Site with competing subsurface usage (e.g., tunnels) Outside settlements Unit susceptible for heavy karstification (St. Ursanne Fm., Upper Muschelkalk Fm.) Geological unit with the risk of rock swelling and subrosion	Not allowed (A)
Groundwater protections area A_U Area with the risk of karstification Area with multiple aquifer levels, confined artesian aquifers, saline aquifers Area with geogenic risks (landslides, oil shale, natural gas, rock swelling, subrosion) Area with insufficient geological information Capture zone of mineral or thermal springs	Allowed with specific requirements (B)
Other area	Allowed with standard requirements (C)

examples including karst systems, areas with the threat of rock swelling and subrosion, water protections zones, and areas with multiple aquifer levels or confined artesian aquifers. The examples aim at illustrating the complexity that has to be considered when implementing geothermal systems in complex urban geological environments. It is emphasized that local impacts can have severe effects on large areas at the regional scale.

5.5.2.1 Karst Systems

Karst systems are extremely heterogeneous and groundwater flow is complex (see Sect. 5.4). Therefore, karst areas are specifically problematic with respect to BHE drillings. Problems are associated with the loss of drilling fluid, with borehole stability and with the backfilling of the borehole, and possibly the collapse of cavities. In particular, the groundwater availability and quality is endangered by uncontrolled losses of drilling liquid in the subsurface, which may block existing flow paths or cause far-reaching groundwater contamination. In addition, new permanent pathways can be generated by the drilling, causing a change in hydraulic conditions. Because of these processes, the risk of groundwater contamination by BHE drillings is appraised to be specifically high in karst areas. To meet this risk, a specific proceeding for the installation of BHE systems in karst areas is required as shown in Fig. 5.32. Both the actual situation at a certain site (existence of cavities) and civil planning aspects (downstream drinking water usage) may be considered.

In addition, it is possible to abandon the installation of BHE systems in geological formations that are known for frequent and heavy karstification. In Basel-Landschaft,

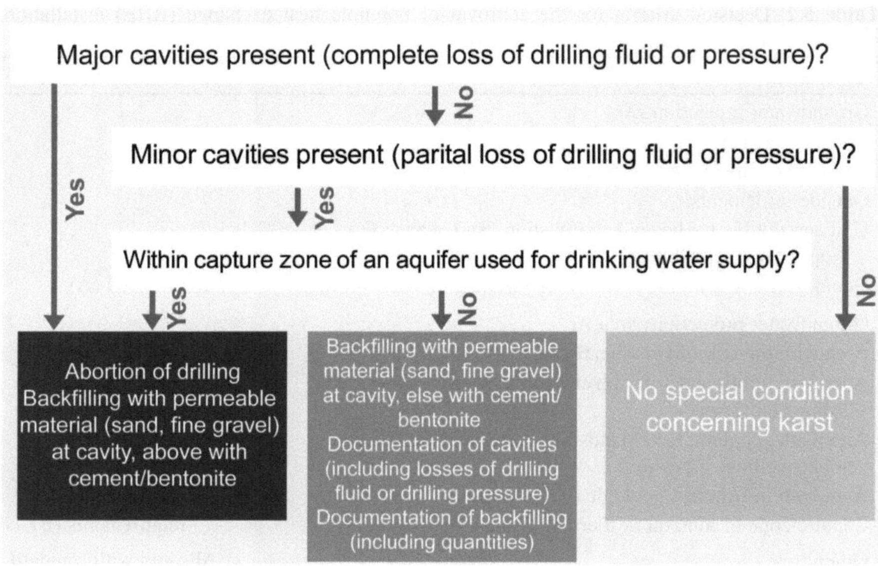

Fig. 5.32 Flow chart for the proceeding of drillings made for the installation of borehole heat exchangers in karst areas

for example, the Jurassic (Oxfordian) St. Ursanne Formation as well as the Triassic Upper Muschelkalk Formation is specifically susceptible for heavy karstification, and therefore inappropriate for BHE installations.

5.5.2.2 Areas with the Threat of Rock Swelling and Subrosion

The drilling of boreholes for BHE installations weakens the rock around the borehole and creates a zone with increased rock permeability. Moreover, the backfilling of the borehole may be incomplete. These features may create new pathways for groundwater flow and change the hydraulic conditions in the subsurface.

In areas with sulfate bearing geological units, water access can trigger the transformation of the mineral anhydrite into gypsum. This reaction is associated with a considerable volume increase, resulting in high swelling deformation or pressure. In extreme cases, swelling can result in a floor heave at the land surface. A highly topical example is the town of Staufen in South Germany. The historic center of this town was dramatically affected after the installation of a BHE field in the sulfatic Gipskeuper Formation. More than 250 houses have been seriously damaged (Fig. 5.33). It is known that swelling often occurs when different factors interplay at particular locations, like changes in groundwater flow regimes in combination with the weathering of the liner material. Such phenomena are also known from road building and tunneling. Much less known, in contrast, are the swelling processes taking place in the subsurface, which are related to induced

Fig. 5.33 In the historic center of Staufen, more than 250 houses have been damaged because of swelling ground after the installation of a borehole heat exchange system in sulfatic clay rocks

changes of the hydraulic systems. For this reason, it is not recommendable to install BHE systems within sulfatic geological formations.

The access of water to gypsum or salt containing layers may cause leaching of these compounds and produce gypsum or salt karst with the consequence of subrosion, accompanied by land-subsidence at the surface. Similar to the threat of rock swelling, also at sites with gypsum and salt containing geological units the drilling into these layers has to be avoided. In the urban area of Muttenz-Pratteln close to the city of Basel, subrosion is known from the Anhydrite Member of the Middle Muschkalk Formation (see Sect. 5.4). The observed subsidence, however, is not triggered by the installation of BHE in that case. Potential for subrosion is also assigned to the Trisassic Gipskeuper Formation. Both the Middle Muschelkalk and the Gipskeuper Formation are also candidates for rock swelling.

The underlying strata of the Gipskeuper Formation consist of a few meters of the Lettenkeuper Formation, followed by the Upper Muschelkalk Formation. The latter is a regional aquifer. The groundwater within this aquifer is often confined because the strata in the hanging wall mainly consist of low permeable marlstone. When drilling through the Keuper strata into the Muschelkalk Formation, water flow along the borehole into the sulfatic Gipskeuper layers is difficult to avoid, producing the risk of rock swelling or subrosion. However, even when the depth of the borehole is limited to the Gipskeuper Formation, such swelling phenomena cannot be absolutely excluded. The assumption that drilling into sulfatic geological formations is safe as long as they are not connected with an aquifer by the drilling is too shortsighted. Low permeable aquitard formations are not necessarily "dry" but may contain formation waters. The processes leading to rock swelling are yet poorly understood. When the hydraulic potential is changed by a BHE drilling, the consequences are difficult to foresee. If rocks that are susceptible to swelling or subrosion are penetrated by a BHE drilling, it is difficult – or even impossible – to judge if water access can be avoided. A precondition for reliable predictions would be the accurate knowledge of regional and local hydrogeological constraints. This knowledge can only be deduced from monitoring systems that continuously measure hydraulic parameters and the hydraulic changes induced by the drilling.

5.5.2.3 Water Protection Zone (A_U)

The Swiss water protection area A_U (abbreviation A_U: "A"-areas are inside the protection area, contrasting to "B"-areas; u stands for underground waters) comprises areas with groundwater that can be used for drinking water supply, as well as adjacent areas with no or marginal groundwater occurrence that are necessary to protect the usable groundwater. The groundwater protection area A_U is independent of an actual usage of the groundwater. It suits the purpose of comprehensive and sustainable protection of groundwater resources.

During the installation and operation of BHE systems within A_U, there is a risk of groundwater contamination (1) if drilling liquid or heat transfer liquid of the BHE leaks into the aquifer, and (2) if the protective function of the soil above the groundwater is reduced by the drilling. A permanent connection between the groundwater and the land surface may be created, which provides the chance of contaminant access into the groundwater.

The water protection area A_U is commonly located in river valleys, as are most urban areas. For this reason, the majority of geothermal energy projects using BHE systems could not be realized if such systems were excluded from these areas. However, a safe installation of BHE is feasible if certain measures are taken to minimize the risk of groundwater contamination. These measures include special requirements concerning the heat transfer liquid (some liquids are more hazardous than others) and the borehole installation. The installation has to be designed to separate the borehole section within the aquifer from the section in the underlying aquiclude. This can be achieved by a permanent casing of the borehole from the land surface down to the aquifer basis and the usage of textile borehole packers. Impacts of shallow geothermal installations on the environment may also arise from changes in groundwater temperature, for example by affecting aquatic habitats (Ferguson 2009). In the case of large facilities (BHE fields), calculations concerning the thermal effects on the groundwater are therefore required in addition (e.g., Hähnlein et al. 2010; Pannike et al. 2006).

5.5.2.4 Multiple Aquifer Levels

Commonly, only shallow aquifers are used for drinking water. The groundwater quality of deep aquifers is mostly unknown. There is an increased chance of saline groundwater in deep aquifers. In addition, highly mineralized groundwater from deep aquifers may be corrosive to the backfilling of boreholes, endangering the backfilling's tightness. There is a risk of connecting deep aquifers with shallow aquifers by a BHE drilling, which may decrease the drinking water quality of the shallow aquifer. Moreover, the connection of different aquifer levels alters the subsurface hydraulic system as a whole. The consequences of such alterations are difficult to foresee. Apart from groundwater quality degradation by saline groundwater intrusions into a shallow aquifer, there is the risk of contaminant mobilization in the subsurface by altered flow systems. Changes in flow patterns may also reduce the yield of drinking water wells, and enhance processes favoring rock swelling and subrosion. All these risks suggest that a connection of separate groundwater levels by BHE drillings must be avoided.

5.5.2.5 Confined Artesian Aquifers

The outflow of artesian groundwater may cause contamination and damage at the surface. Often, the water flowing out from the borehole is accompanied by drilling liquid or by the cement used to seal the borehole. The drilling into "only confined" (not artesian) groundwater often remains unnoticed. However, also this can be problematic. The places the water flows to, and the resulting consequences, are commonly not evident.

If there is a pronounced topography, the risk of meeting artesian groundwater during the drilling of a borehole is especially high in valleys, because the hydraulic head is increasing with depth in such situations. If an aquifer is confined, the existence of artesian groundwater must always be anticipated when drilling in a valley. Therefore, in areas with confined groundwater the depth of a borehole must be limited in the way that the aquifer is not reached by the drilling. This requires an accurate knowledge of the spatial configuration of the subsurface layers as well as local and regional circulation systems. A prerequisite is sufficient data availability.

5.5.2.6 Data Requirements

To assess if the criteria that allow or deny an installation of BHE systems are met at a certain location, both land use planning data (e.g., groundwater protection zones) and geological–hydrogeological data (e.g., position/thickness of karst units) are required. The land use data are often readily available at public authorities occupied with spatial planning (see Sect. 4.1). In contrast, geological and hydrogeological data often have to be evaluated and interpreted. If the succession of geological units

Table 5.3 Geological units of the Keuper in the region of Basel and relation to the geothermal use concept of Canton Basel-Landschaft. The installation of borehole heat exchangers within a geological unit is either not allowed (case A, dark gray), or allowed with specific requirements (case B, light gray), or allowed with standard requirements (case C, medium gray)

System-Series	Formation	Typical thickness [m]	Case	Criteria
	Rhät	0–5	(C)	
	Obere Bunte Mergel Fm.	10–40	(C)	
	Gansinger Dolomite	5–15	(C)	
	Untere Bunte Mergel Fm.	0–15	(C)	
	Schilfsandstein Fm.	0–20	(B)	Potential for several aquifer levels, confined aquifer, saline aquifer
	Gipskeuper Fm.	60–150	(A)	Geological unit with the risk of rock swelling and subrosion
Triassic – Keuper	Lettenkohle Fm.	0–10	(C)	

in the subsurface is known, these units can be related to the geological–hydrogeological criteria defined by the geothermal use concept outlined above. Table 5.3 illustrates the relation between geological units of a region and the individual criteria of the presented geothermal use concept exemplarily for the Triassic Keuper Formation. The installation of BHE in units which are assigned to the case (A) is not allowed. These layers are indicated in dark gray. The installation of BHE in units assigned to the case (B) (medium gray) will need the conformance to special requirements. If the succession and thickness of the subsurface units in the subsurface are known, a maximum borehole depth can be determined that allows the installation of BHE with special or standard requirements. Such information can be implemented into geothermal energy use maps using a Geographic Information System (GIS). The GIS may include maps showing areas where BHE installations are (or are not) allowed, and indicate limitations (e.g., maximum drilling depth) that apply to BHE projects in certain areas. Both the spatial configuration of land planning data (e.g., groundwater protection zones) and hydrogeological information (e.g., spatial distribution of karst areas) can be provided by such GIS-based geothermal energy use maps.

5.5.3 Application of Monitoring and Modeling Methods

In the following case study, we show for selected areas in the Canton Basel-Stadt how geothermal energy use concepts can be implemented (Epting et al. 2010). We focus on how to accommodate existing and future groundwater use to long-term goals and a sustainable development of urban groundwater resources. Monitoring and modeling approaches focus on the quantification of the thermal use potential within the unconsolidated sediments in northwestern Basel. The study starts with a characterization of the present thermal state of the groundwater in the city of Basel, as a base for defining short-, medium-, and long-term goals for the development of the groundwater quality. In the city of Basel a heating network already exists, therefore groundwater is more frequently used for cooling purposes.

The upgrading of regenerative heat and energy sources requires process knowledge, based on a combination of field experiments, monitoring systems, and numerical models that couple flow and heat transport. This enabled us to simulate regional long-term heat transport and to quantify possible developments of temperature distributions in urban aquifer systems.

Groundwater-sourced heating or cooling typically results in a thermal plume of cool or warm reinjected groundwater and can cause regional changes in groundwater temperatures. Such a plume may be regarded as anthropogenic thermal pollution. In particular, increasing geothermal groundwater use can exceed the subsurface potential for different heating and cooling systems and thus affect groundwater quality. Thermal groundwater use and the reinjection of water with elevated or decreased temperatures inevitably leave a fingerprint on the aquifer. Questions arise, as: To what extent can the fingerprint influence downstream groundwater use or how pronounced is the memory effect of the aquifer when considering life-spans of thermal groundwater use?

The illustrated investigation methods for shallow geothermal energy use focus on:

1. The development of state-of-the-art devices for depth-oriented temperature monitoring.
2. Numerical heat flow and transport modeling of shallow geothermal systems.
3. Optimization of existing and new geothermal systems.

5.5.3.1 Depth-Oriented Temperature Monitoring

For the set up of heat transport models, the thermal stratification within the aquifer should be calibrated with vertical temperature profiles. Four state-of-the-art observation wells for depth-oriented temperature monitoring (Fig. 5.34) focus on capturing several spatiotemporally extremely heterogeneous processes, as:

1. River–groundwater interaction effects on the thermal regime near the river.
2. Temperature stratification downstream of thermal groundwater use.
3. Effects of heated construction-parts reaching into aquifers and their downstream influence on groundwater resources.
4. Depth-oriented temperature monitoring at the southern (inflow) model boundary.

Capturing these processes and boundaries is important to assess, quantify, and qualify the risk of groundwater contamination by thermal groundwater use.

Results of the depth-oriented temperature monitoring also improved the interpretation of existing temperature measurements within the investigation area. A preliminary statistical analysis of the existing temperature time series revealed

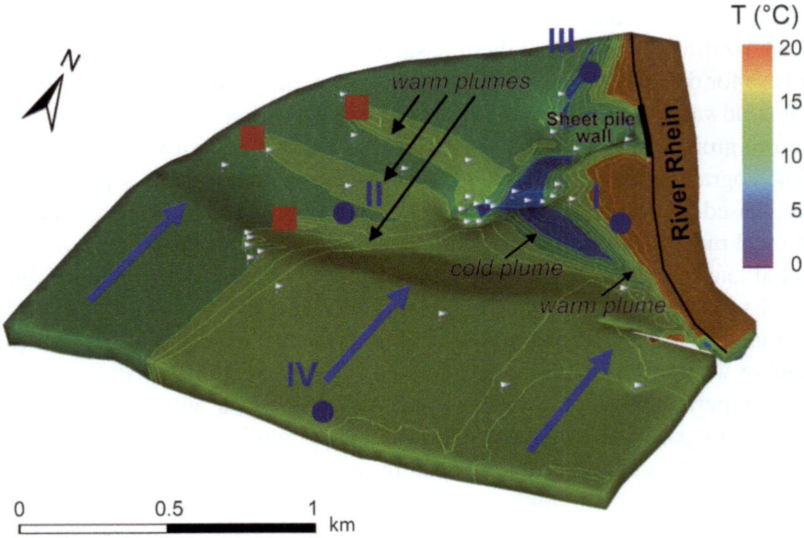

Fig. 5.34 Simulated groundwater temperatures for northwestern Basel and locations of observation wells (*dots*) as well as reinjections (*rectangles*). *Blue arrows* show the regional groundwater flow direction. Locations I–IV of observation wells for depth-oriented groundwater temperature measurements

missing information making data interpretation difficult. The missing information is related to:

1. The vertical position of the temperature sensor within the observation well.
2. The position of the observation wells' screen within the flow field.
3. The consideration of the infiltration of surface waters.
4. The sensitivity concerning changes within the regional groundwater flow regime.
5. Groundwater usage.
6. Possible measurement errors and data jumps.

Figure 5.35 illustrates first results of the depth-oriented temperature monitoring for observation well (I) close to the river. High temperatures in the saturated and nonsaturated zone are attributed to the heating period in winter. Furthermore, we hypothesize that during flood events in summer "warm" water from the river infiltrates and stratifies in the upper part of the aquifer. During flood events in winter "cold" water from the river infiltrates and stratifies in the lower part of the aquifer. Overall the preferential thermal propagation in more conductive river deposits can be observed.

5.5.3.2 Numerical Heat Flow and Transport Modeling

The increasing use of regenerative heat and energy sources increasingly requires the application of models, which enable an integrated consideration of all relevant hydrological and hydrogeological processes, technical interferences, and existing

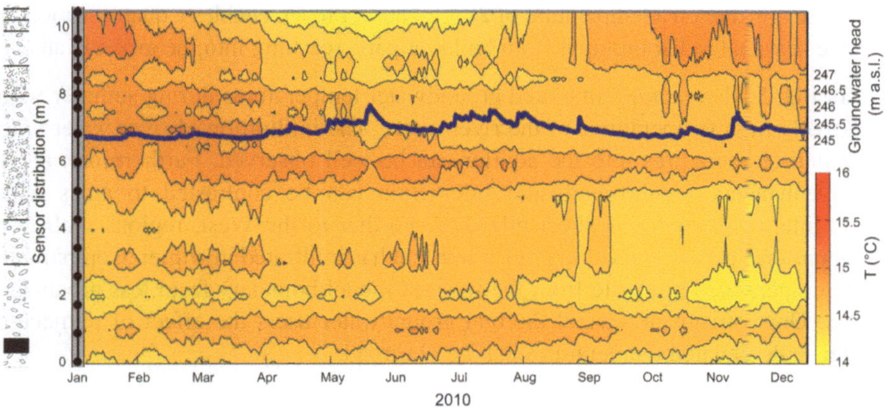

Fig. 5.35 Observation well for depth-oriented groundwater temperature measurements. Temperature isolines for observation well I; left: lithostratigaphic borehole information (see Fig. 5.34 for location)

utilizations. Groundwater models facilitate the simulation of regional long-term heat transport and quantification of temperature changes (Cathomen et al. 2002). Fujii et al. (2007) proposed the development of suitability maps for ground-coupled heat pump systems using groundwater and heat transport models.

We performed numerical simulations with a groundwater model covering the area of northwestern Basel. For the setup of the model geometry and hydraulic boundary conditions, please refer to Sect. 5.2. The model was extended for groundwater heat transport simulations using the groundwater modeling software FEFLOW. First calibrations were performed with existing head (17) and temperature (11) time series for the time period 01.04.2006 to 31.03.2007. Results enabled us to define the current status of the hydraulic and thermal groundwater regime within the investigated area, including the evaluation of seasonal temperature patterns. Subsequently, we performed fine-tuning with time series from depth-oriented temperature monitoring for the time period 02.02.2010 to 10.01.2011.

For the regional thermal groundwater regime, we assumed that based on natural factors that influence groundwater temperature a characteristic yearly average temperature develops. Anthropogenic boundary conditions as caused by groundwater extractions and injections change these average conditions. Temperature gradients develop that induce thermal flow. Three thermal sources and sinks are considered as boundary conditions and partially had to be estimated:

1. Influence of precipitation water and areal groundwater recharge on groundwater temperatures.
2. Injection of water that was used for cooling (for the estimation of thermal input of injected water an average heating of the injected water of 4°C was assumed).
3. The yearly variable radiation of construction parts reaching into the groundwater (cellars, underground parking lots, channelization and systems for long-distance heating, etc.). For the thermal input of heated cellars it is assumed that

these are kept on a temperature of 21°C. This approach enables a quantification of the seasonal radiation input from building parts reaching into the groundwater.

In Fig. 5.34 we show simulated groundwater temperatures during summer with relatively high temperatures of the river water. The infiltration of river water with elevated temperatures into the aquifer can be well observed. Furthermore, areas with enhanced river–groundwater interaction can be contrasted to areas with little interaction due to a sheet pile wall. Further to the West, regions with low temperatures derive from cold river water that has infiltrated in winter. Comparable low temperatures originate from the western model boundary from less urbanized areas. Downstream of several thermal groundwater users the effect of reinjected water with elevated temperatures can be observed.

5.5.3.3 Optimization of Geothermal Systems

With the calibrated model previously defined scenarios enabled us to evaluate temperature changes within the local and regional groundwater flow regimes that come along with new thermal groundwater use. The results of the process-based investigations are the basis for dimensioning and for site evaluation of new thermal groundwater facilities, and enables the optimization of operation schedules as well as extraction and injection locations. Furthermore, the environmental impact and potential hazards on groundwater resources can be evaluated.

The various scenarios for the optimization of future management focus on:

1. Groundwater management programs (temporal optimal organization of extraction and injection of groundwater).
2. New locations for groundwater use (optimal spatial integration into existing supply networks and consideration of subsurface infrastructures) considering long- and medium-term development.
3. Studying long-term effects of climate change on an urban water body, considering positive and negative feedback mechanisms as well as the relevant spatio-temporal scales. Furthermore, for different regions within the investigated area residence times of thermal changes and long-term shutdown of geothermal systems (“memory effect” of the subsurface) are evaluated.

5.5.4 Conclusions

For the sustainable development of urban subsurface resources, adequate management concepts are required. This includes the setup of tools that enable the investigation of the relevant processes that dominate groundwater flow and thermal regimes at different spatiotemporal scales.

A sustainable usage of shallow geothermal systems requires knowledge of the hydrogeological characteristics of the subsurface and an understanding of the processes. The regulations concerning the application of the technology must be based on this knowledge and process understanding. Differentiated legal regulations can support the advancement of low temperature geothermal energy and, at the same time, minimize negative impacts of geothermal systems on the environment. Criteria for decisions related to the authorization of geothermal installations have to consider not only constraints given by land use planning, but also the specific geological characteristics of a region, comprising different aquifers and, for example, formations being susceptible to subrosion or swelling.

In order to reduce negative impacts on the environment by the use of low temperature geothermal energy, environmental regulations regarding the installation and operation of geothermal energy systems are required. However, the implementation of legal requirements into practical authorization may reflect a liberal or restrictive attitude of the responsible public authorities towards the installation of geothermal systems. The authorities have to trade off the benefits against the risks, both for the individual and for the public. If they want to advance the renewable energy provided by low temperature geothermal systems in a sustainable way, they have to consider the risks associated with the according technology.

The public geological surveys play an important role in the sustainable advancement of shallow geothermal systems. Generally, they support the planning of geothermal installations by providing geological information. The more specifically the geological data are processed, and the easier they are made available, the more beneficiary they are. By providing adequate planning tools, the public geological surveys can advance both the use of geothermal energy and the protection of the environment. As subsurface processes are a 3D problem, planning tools should be based on 3D geological models. They may also include GIS-supported geothermal maps and numerical simulations of groundwater flow and heat transport. Ideally, the access to such applications would be web-based.

We presented an approach, where data from high-resolution depth-oriented temperature monitoring are incorporated into 3D numerical models that facilitate the simulation of groundwater flow and thermal regimes. The results form the basis for the development of concepts for the use of the urban subsurface for specific aquifer regions and future use scenarios.

While consequences of climate change are difficult to detect in groundwater resources and remediation requires global strategies, regional and local impacts are often orders of magnitude larger, in particular in urban areas. The consequences of these effects can however be minimized by optimizing water resource management.

5.6 Natural Hazards in Urban Areas

Peter Huggenberger and Donat Fäh

According to Dölemeyer (2009), natural hazards such as flooding or earthquakes are first and foremost natural phenomena which do not necessary lead to disasters. They are part of nature. To turn into a disaster, the event has to hit a vulnerable human society, e.g., an urban area, which is susceptible to its damaging effects.

The knowledge of processes that are related to natural hazards is of societal relevance especially for urban areas. In this section, we illustrate how to assess natural hazards in the urban area of Basel. We focus on hazard identification and a discussion on the associated geological and hydrogeological processes. According to palaeo-seismological research results mainly from the Basel-Reinach fault system and from traces in lake sediments, five to eight 1356-type earthquakes can be identified that occurred during the last 10,000 years (Becker et al. 2000, 2002; Becker and Giardini 2001; Meghraoui et al. 2001; Ferry et al. 2005; Fäh 2009). Thus, the recurrence interval of magnitude 6.5–7.0 events is in the order of about 2,000 years. For flooding there is indirect evidence for extraordinary events at the same frequency as for earthquakes. The flood of 13 June 1876 is considered to be one of the largest historical floods documented in the history of Basel. For the 1876 event the discharge of the river Rhine was in the order of 5,800 $m^3\ s^{-1}$ (Pfister 1999). However, based on the sedimentological record, there is evidence for larger flood events during the last 12,000 years exceeding the size of the 1876 event considerably (Kock et al. 2009). Such events might have had landscape shaping capacities. It is known that the present landscape in alluvial valleys, such as the Upper Rhine Graben, is the product of endogenic (earthquakes or regional tectonic deformation such as uplift or subsidence) and exogenic processes (flooding and erosion) and we might ask how many earthquakes and floods where necessary to create this landscape.

Besides earthquakes and flood events there are a series of smaller scale phenomena such as sink holes, landslides or rock falls, and local subsidence due to subsurface evaporite dissolution (Sect. 5.4). These processes can have some important local impact in urban areas such as on traffic lines or lifelines, but would not lead to regional scale disasters. In this section, we illustrate the impact of earthquakes (natural but also induced triggered events) and catastrophic floods in the Basel area. We also include a short description of current techniques for quantitative earthquake microzonation and the evaluation of catastrophic flood events.

5.6.1 Earthquakes in Urban Areas

In many urban areas with moderate seismic hazard, the consequences of large seismic events have often been underestimated (Fäh et al. 1997). For the region of Basel, located at the southern margin of the Upper Rhine Graben, large earthquakes were considered to be less probable compared to other geohazards such as catastrophic floods and therefore were considered as nonrelevant in urban planning. In the 1980s, scientist of the Swiss Seismological Service started with the analysis of the damage caused by the historical earthquake in 1356 AD. This event is considered to be the largest historical event yet known in Central Europe (Mayer Rosa and Cadiot 1979). Recent studies assessed the damage potential of an event of the size of the 1356 earthquake by developing damage scenarios taking into account the urban development in the past centuries from a small city of about 6,000 inhabitants in the fourteenth century to more than 300,000 in the wider agglomeration of Basel today (Fäh et al. 2001, 2009a). It is now accepted that the financial consequences of a large earthquake in the region of Basel with a recurrence of about 2,000 years are in the order of 50 billion Swiss Francs. Because of the large damage potential and the high financial impact, such earthquakes have to be considered with attention by developing adequate mitigation strategies including building codes, land-use planning, civil protection measures, and economical aspects. Even small earthquakes (intensities V to VI) can cause considerable economic impact as was the case for the induced events in 2006/2007 (Geopower Basel 2007).

Man-made triggered earthquakes were a major political and societal issue after the 2006/2007 series of earthquakes within the course of the development of an enhanced geothermal reservoir at a depth of about 5 km underneath the city of Basel. The largest event widely felt in the Basel area reached a Richter magnitude $Ml = 3.4$. It was triggered on 8 December 2006, after suspending the stimulation of the reservoir in the morning of the same day. As a consequence the project was suspended and finally stopped. The seismic activity however continued and several events were felt by the population over the period of several months, all of them induced by this Deep Heat Mining (DHM) project. An extensive study, commissioned by the Canton Basel-Stadt and supported by the Swiss federal government, assessed the seismic risk resulting from a continued development and subsequent operation of a DHM geothermal system in Basel. This study concluded that the risk of a geothermal project to cause bodily harm is low but the property damage may be deemed as unacceptable according to the risk criteria of the Swiss ordinance on major accidents (Baisch et al. 2009). A comparison with other technical risks in Switzerland came to the same conclusion (Baisch et al. 2009). These authors also concluded that from the seismic risk perspective, the location of Basel is unfavorable for the exploitation of a deep geothermal reservoir in the crystalline basement with stimulated DHM techniques. However, all techniques that lead to changes in pore pressures lead to changes in effective stress, which is responsible for deformation, and might induce earthquakes.

The basis for a risk study is the assessment of the seismic hazard in the area. Even if the seismic hazard is moderate, the seismic risk in Basel is relatively high due to the dense urbanization and important infrastructure. In order to define the level of seismic action on terms of ground motion to be expected for a given return period, three factors have to taken into account:

1. The characteristic of the seismic sources relevant for the area (location and size distribution, depth, rupture properties, and direction of emitted energy).
2. The length of the propagation path of the seismic waves and the related attenuation properties of the crustal structure between the source and the site.
3. The local geological conditions that can amplify or deamplify seismic waves and can increase the duration of seismic ground motion due to resonance and locally excited surface waves.

The quantitative assessment of these three aspects is the goal of local seismic hazard assessment, called microzonation. In most cases, the quantification of seismic hazard is given in terms of a probabilistic description of expected ground motion (e.g., 10% probability of exceedance in 50 years equivalent to 475 years return period). Such hazard products are implemented in modern building codes, and expected to be used by structural engineers to design new buildings. Microzonation is a tool to map areas of elevated hazard related to ground motion as well as of possible triggered phenomena, such as liquefaction, triggering of landslides and rockfall, and ground settlement triggered by earthquake ground motion.

5.6.1.1 Earthquake Microzonation

Earthquakes show some differences compared to other natural hazards (1) they might create damages on vast areas; (2) they occur without warning; and (3) they cannot be stopped. Until today there are no reliable methods to the time, the locality, and the intensity of earthquakes with sufficient precision. Earthquakes worldwide have a massive destruction power on buildings, infrastructures, and industrial plants as recently demonstrated by two relatively small earthquake around magnitude 6, a typical size for Swiss earthquakes of the past (Fäh et al. 2003; Wiemer et al. 2008). The first example is the Mw 6.3 Christchurch New Zealand event of 22 February 2011 causing widespread damage and multiple fatalities, followed nearly 6 months after the 7.1 magnitude 2010 Canterbury earthquake that caused significant damage to the region but no direct fatalities. The second event is the 6 April 2009, Mw 6.3 (Ml 5.8) L'Aquila central Italy earthquake during which 308 people are known to have died, making this the deadliest Italian event since the 1980 Irpinia earthquake. These earthquakes have in common that the fault was at shallow depth and very close or within the urbanized area. The same has to be expected for future events in the Basel region due to the presumed location of the large events of the past such as the Mw 6.6 Basel event in 1356 (e.g., Ferry et al. 2005).

The analyses of the damage generally show that there are buildings that sustain an earthquake just beside buildings of the same type which collapsed (Fig. 5.36). This might be explained by the interaction of the seismic response on the particular geological structure and the behavior of the construction during seismic wave excitation. Damage may be induced by amplification of seismic waves due to the geomorphologic settings (i.e., river terraces, sedimentary basins), liquefaction of water saturated soils, as well as due to landslides and rock-fall.

The main reason of the risk potential of earthquakes is the fact that for a long time no adequate precautionary measures were undertaken. For Switzerland, for example, adequate building regulations only exist since 1989, which often are only partially implemented. With an adequate preparation, even when considering earthquakes above the design level of building codes, the earthquake impact on buildings can be reduced.

Fig. 5.36 Kobe earthquake 1995 illustrating that there are buildings that sustain an earthquake just beside buildings of the same type which collapsed. Copyright (C) Free Software Foundation, Inc.

Due to the fact that the damage of earthquakes is related to the shaking of the subsurface and the building itself, as well as their interaction, scientist started to implement microzonation as a base for engineers to optimize the design for the buildings (Fig. 5.37). Target-oriented constructional measures that are optimized in relation to the local geology can result in a reduction of damage caused by earthquakes. Such procedures are of particular significance when old buildings need cost-efficient retrofitting. Microzonation is mapping the geologic influence on earthquake action and the resulting ground motion at given return periods. The goal of earthquake microzonation is also to identify zones with unfavorable behavior in order to use this information in urban planning and for designing buildings in an adequate site-specific manner.

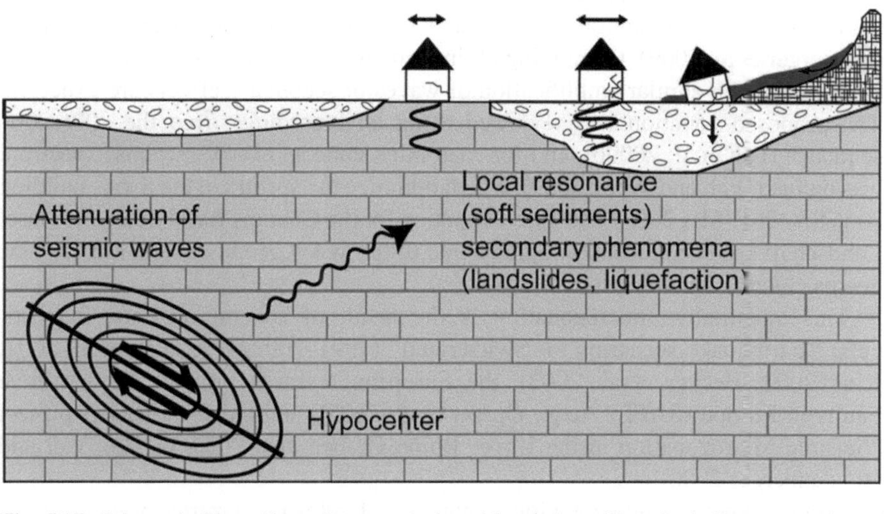

Fig. 5.37 Schematic illustration of source, path and local sites effects including attenuation of seismic waves, local resonance due to soft sediments, and situations with secondary phenomena such as landslides and liquefaction

The region of Basel is densely populated with extensive traffic and communication infrastructures as well as sensitive and security relevant technical facilities. For this reasons, Fäh et al. (1997) performed the first qualitative microzonation for the city of Basel in order to map the quality of soil deposits and to provide an instrument for the Security Department of the city of Basel that was tasked to develop a strategy for retrofitting vulnerable cantonal buildings and lifelines. The zonation was mainly based on the properties of the Quaternary sediments, the influence of the groundwater table on liquefaction, the lithology of the base of unconsolidated rock or deep sediment, and the distance to the eastern master fault of the Rhine Graben, a geologic transition which can generate damaging local surface waves. This study showed that the deep sediments of the Rhine Graben significantly contribute to site effects by inducing a fundamental resonance at very low frequencies (0.3–1.3 Hz) that is accompanied by amplification of seismic waves over a broad frequency band relevant to buildings. Resonance phenomena, edge generated surface waves, and local focusing of seismic waves can induce an elevated level of ground motion in the basin present in the Upper Rhine Graben area. As there was not sufficient earthquake recordings and derived amplification data available, the result was limited to a qualitative assessment of the ground motion.

According to the complexity of the possible effects, a quantitative spectral microzonation study was proposed for the Basel area. The project started in 1997 for the city of Basel, and the area of interest was extended in 2003 into areas in Canton Basel Landschaft and a small region in Canton Solothurn. The procedure includes a 3D mapping of the subsurface geological structure (Zechner et al. 2001), an evaluation of the behavior of the local ground through geophysical measurements (Kind 2002; Kind et al. 2005; Havenith et al. 2007) and numerical modeling (Kind 2002; Oprsal et al. 2005; Oprsal and Fäh 2007; Fäh and Huggenberger 2006), the deployment of seismic instruments in temporary networks and permanent modern strong motion stations, and the analysis of earthquake recordings (Ripperger et al. 2009). Combining all information, the area was divided in microzones for which similar amplification of waves or secondary effects are expected. The proposed zonation was reviewed using all recordings from the 2006/2007 sequence (Fäh et al. 2009b), and for each microzone an elastic response spectrum was defined (Fäh and Wenk 2009), which optimize the spectra in the Swiss building code SIA 261 (SIA 2003). The microzonation for the Cantons Basel Stadt and Basel Landschaft has been recently implemented (http://www.geo.bl.ch/ and http://www.geo.bs.ch/erdbebenmikrozonierung).

This quantitative microzonation is the result of several projects, in particular "Earthquake scenarios for Switzerland" (1997–2002) (e.g., Fäh et al. 2001; Kind 2002; Becker et al. 2002), the European research project "Slow Active Faults in Europe (SAFE)" (e.g., Oprsal et al. 2005), and the INTERREG project "Seismic Microzonation in the Upper Rhine Graben area" (2003–2006; Fäh and Huggenberger 2006).

Along with the microzonation a number of relevant hazards products were established. Maps with geological information (e.g., surface geology, bedrock depth and composition, groundwater table) and tectonics of the area were derived.

Areas of potential secondary effects such as landslides or settlements were mapped (see Box in Chap. 2). A large number of ambient noise measurements were performed in order to derive a map of the fundamental frequency of resonance f_0 of the ground from single station recordings, using horizontal-to-vertical (H/V) spectral ratios or wave polarization analysis (e.g. Fäh et al. 2001). The frequency about half the f_0 on one hand indicates the lower boundary of the frequency range where generally ground motion amplification occurs. On the other hand f_0 is linked through the average shear-wave velocity to the depth of the impedance contrast responsible for strong resonance effects that generally occur at frequencies around f_0. Within the structure of the Rhine Graben the fundamental frequencies are very low (0.4–1 Hz) as a result of soft sediments that can be found down to large depths. Fundamental frequencies outside of the Rhine Graben are generally higher, above 2 Hz. The amplitude of the H/V measurements indicates the contrast in velocities between rock and unconsolidated sediments (Fig. 5.38). Large amplitudes within the H/V spectrum (good to very good quality of the amplitude) indicate a high contrast which can lead to significant amplification due to resonance behavior of the unconsolidated sediments.

A geophysical model (parameterized geometry) was constructed from a geological 3D model by assigning a seismic velocity model to the different geological units from measurements and collected literature values. S-wave velocities were determined using a large number of ambient vibration array measurements as well as active seismic methods (Kind et al. 2005; Havenith et al. 2007). The geophysical model was verified against the measured fundamental frequencies, which showed that the thickness of the sediments with low S-wave velocities in the deep Rhine Graben had been underestimated in many parts of initial model geometry. The measured fundamental frequencies and wave velocities, the known total depth of geological units from boreholes, and the characteristics of the soft sediments were used to optimize the model of the project area. The resulting geological/geophysical 3D model of the Basel region (see Sect. 4.2) allowed us to derive the depth of the Quaternary sediments, the depth of bedrock, and the layering at specific locations.

In a next step, site amplification of seismic waves was calculated on the basis of 2D sections of the full geophysical model (Kind 2002), as well as for the entire 3D model of the deep Rhine Graben structure (Oprsal et al. 2005) and a small region outside the Rhine Graben with the particular gravel terraces typically found in this area (Fäh et al. 2006; Oprsal and Fäh 2007). The modeling also included a large number of 1D numerical simulations to particularly model the variability of the response of the shallow surface layers (Fäh and Huggenberger 2006). Together with the map of fundamental frequencies, the amplitudes of the H/V spectral ratios, the geological and geophysical model, the tectonic information, and the observed amplifications from earthquake recordings, a zonation of the study area was proposed.

We divided the Basel area in microzones defined by similar low-frequency response, whereas each of the microzones is again subdivided into subzones defined by the characteristic behavior of the shallow layers that characterize the high-frequency response (Fig. 5.39). Within one microzone three geological subzones can be defined (Loess and clay, unconsolidated Holocene deposits, old deposits).

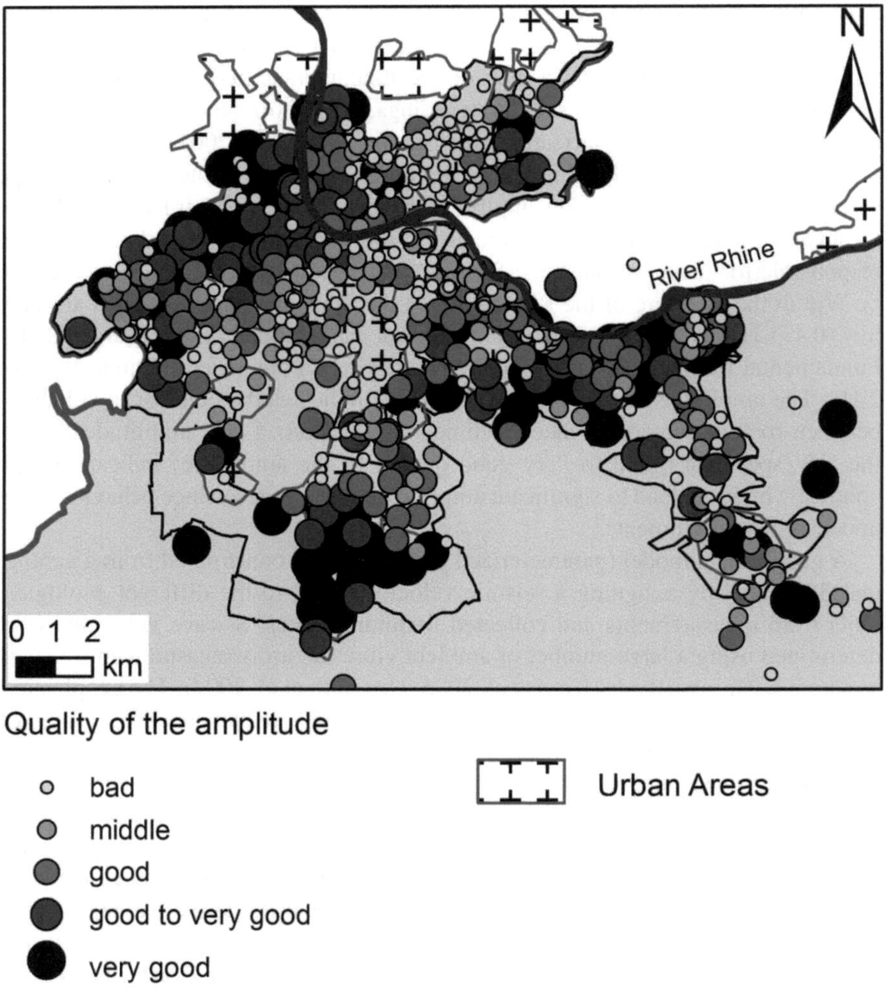

Fig. 5.38 Polarisation (H/V amplitude) at the measured fundamental frequency of resonance indicating the impedance contrast between sediments and bedrock, and the therefore the possibility of resonances

These kinds of near-surface deposits react differently during earthquakes as they are characterized by different S-wave velocities (Fig. 5.40). Within one subzone a similar response during earthquakes is expected. For these subzones we then defined an elastic response spectrum that allows engineers to optimize the design of the buildings to the particular site (Fäh and Wenk 2009). A reference rock response spectrum was determined for each microzone based on the probabilistic regional hazard for the usual return period of 475 years for normal buildings, and was combined with the expected amplification in the subzone.

The resulting spectral values for the horizontal acceleration are given for a return period of 475 years in accordance with the new Swiss building code that is based on

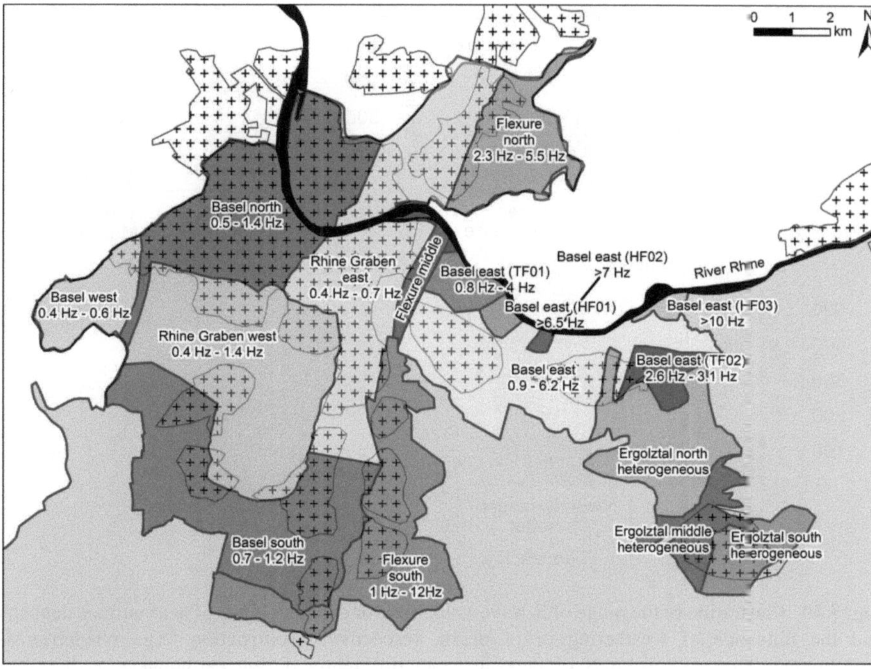

Fig. 5.39 Division of the Basel area in microzones defined by similar low-frequency response, whereas each of the microzones is again subdivided into subzones defined by the characteristic behavior of the shallow layers that characterize the high-frequency response. The range of the fundamental frequency of resonance is provided for each microzone

the Eurocodes. To avoid inefficient designs, the seismic action is now given in a more precise manner than conventional response spectra based on soil classes and seismic zones. In a final step, the spectrum obtained from the probabilistic hazard was combined at long periods with the spectrum based on the return period of events of the size of the 1356 Basel earthquake, the strongest historical earthquake known in the area. This procedures guarantees that at least the structures in the highest importance category, i.e., essential large facilities, will resist a seismic event similar to the Basel earthquake. In Fig. 5.41, we illustrate two elastic response spectra as can be found outside and within the Rhine Graben, respectively. The Tertiary deposits in the Rhine Graben (low shear-wave velocities at large depth) result in higher spectral values at lower frequency.

The quantitative microzonation describes the expected ground motion only. However, it is known that so-called triggered or secondary effects frequently occur during seismic events with strong shaking. These effects include landslides, rockfall, liquefaction of water-saturated unconsolidated sediments, and ground settlement of artificial fills which also generally show more amplification of seismic waves. Based on historical events and on the information from GeoData (see Sect. 4.1), the present knowledge of features susceptible to such effects has been

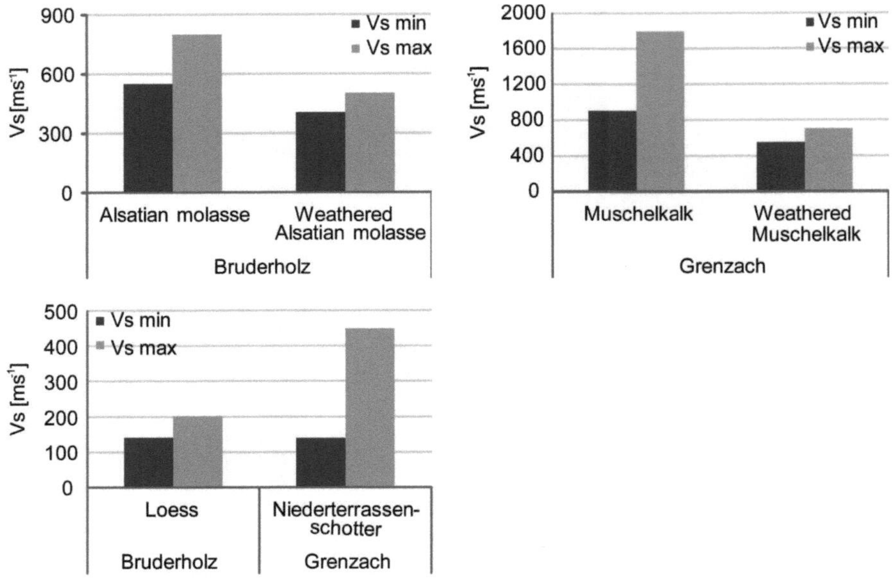

Fig. 5.40 Illustration of the range of S-wave-velocities for different kinds of near-surface deposits and the influence of weathering or of depth, respectively compaction (e.g., Niederterrassenschotter). The various kinds of deposits react differently during earthquakes as they are characterized by different S-wave velocities

summarized on a GIS map. Thereby, the possible triggered phenomena have been translated into an easily accessible GIS product that allows retrieving the specific geological information to be considered in construction projects.

Together with the microzonation map it serves as an information platform for engineers and the public administration. However, it is clear that such maps are never complete but represent the present state of knowledge. The available GIS project includes a definition of the project perimeter, the definition of microzones and subzones, and the response spectra for each zone. In addition, geological information relevant for the seismic shaking as well as possible triggered phenomena such as instable slopes, sink-holes, etc., are integrated. The information can be downloaded from the following public Web sites: http://www.stadtplan.bs.ch/geoviewer/ and http://www.geo.bl.ch/.

5.6.2 *Flood Events in Alluvial Valleys*

Compared to earthquakes, damage due to catastrophic flooding concentrates mostly on the lower most part of an alluvial plain. The climate conditions during the formation of the main valley after the last glaciation were slightly different to those today. However, larger floods that exceed the present knowledge are possible

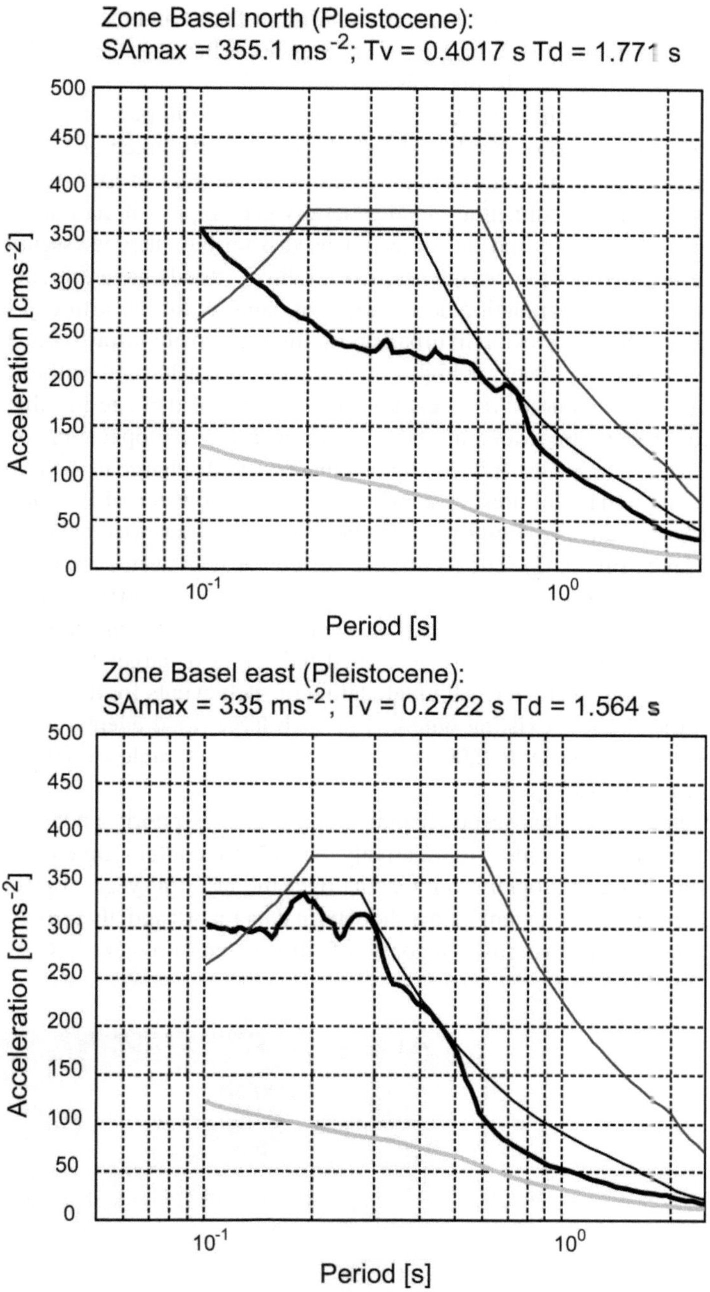

Fig. 5.41 Examples of two elastic response spectra for microzones outside (bottom) and within the Rhine Graben (top), respectively (*thick black curves*). The enveloping *thin black curve* shows the classical response spectra. The parameters for the curve are annotated in the title of the graph. The *thin dark gray line* shows the spectrum of the building code SIA261. The *thick light gray curves* with the lowest values correspond to spectrum for rock. The thick deposits in the Rhine Graben result in a higher values at low frequency

(Mandelbrot and Wallis 1968; Mosher and Martini 2002) and there are sedimentological arguments that the fluvial terrace sequences in the Rhine valley formed by a series of catastrophic floods after 12000 BC.

Due to landscape changes at the catchment scale the impact of floods, such as channel migration, was reduced drastically by water management in the headwaters or storage of water in artificial or naturals lakes. In Central Europe, along regulated streams, most of the sedimentary load generally gets trapped in natural or artificial reservoirs such as lakes and the water released from the lakes is essentially sediment free or poor. As a consequence, extraordinary floods have a great erosional power causing bed degradation downstream of dams as well as artificial or natural bedrock steps. Strong erosion in urban areas might excavate infrastructure installations such as gas pipelines or high power cables, often located within river side dams or in the foreland of river banks. Another type of damage in urban areas is related to river–groundwater interaction, where a rapid propagation of pressure waves in the groundwater, induced by fast increase in discharge in the rivers, might create ground or dam failures. In some cases, the groundwater table rapidly rises up to the surface and is the main reason for flooding of entire towns that are located in alluvial plains (Fig. 5.42). The recurrence periods of such events are in the order of 100 to 300 years. These types of events cannot shape landscapes considerably.

However, within the sedimentary sequences outcropping in many gravel pits along the river Rhine valley, we identified indicators of larger flood events with landscape shaping character. Although dating of such events for the Basel area is a difficult task, the few existing outcrops of such large flood events indicate recurrence periods in the order of 2,000 years, which are comparable with the recurrence periods of earthquakes.

The frequency of landscape shaping flood events is poorly understood as the hydrological data records are relatively short and contain much uncertainty. A relatively good hydrograph only exists over the last 150 years. Paleohydraulic techniques offer a way to lengthen a short-term data record and, therefore, to reduce the uncertainty in hydraulic records of direct measurements (Costa 1987).

Fig. 5.42 Flooding in Laufen. The colored red water indicates the overflow of oil tanks

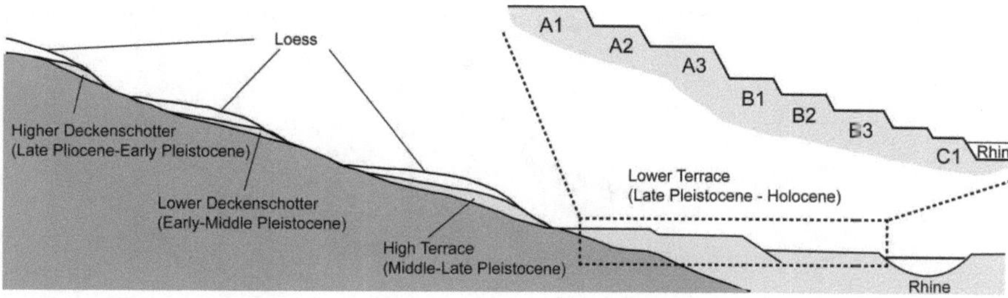

Fig. 5.43 Schematic situation without scale of Pleistocene fluvial deposits between the Aare inlet and Mulhouse. The original terrace surfaces of the Lower Terrace are generally conserved, being only covered by soil, contrary to the older terraces, which are covered by a thick loess layer. The enlargement shows the sublevels classification of Wittmann (1961)

Paleohydraulic data typically have been used to quantitatively reconstruct hydraulic variability for about the last 10,000 years. Beyond 10,000 years, quantitative paleohydraulic investigations often are hindered by limited evidence of channel changes. In addition, in dynamic coarse braided stream environments, sediments are permanently reworked. This makes correlations of the records of lake sediments with the ones of the active flood plain very difficult.

The geomorphologic processes during the last 18,000 years lead to the complex association of fluvial terraces. In the Basel area, a preliminary chronological framework of the age of the river terraces (Fig. 5.43) indicates an age of the Münster Terrace level in the order of 12000 years BC, and lower levels (Pré) Bölling, Early Holocene, Atlantikum, and Subboreal (see Table 2.1, Rentzel 1994).

River terraces are common examples of preserved, geomorphic features. In the literature, two classes of river terraces are typically defined: aggradational and degradational or erosional. Whereas the highest terrace in the Rhine valley has been formed by aggradation, the lower ones are the result of erosion and subsequent degradation, some of them combined with some aggradation (Fig. 5.44). The morphological map of the river terraces clearly shows that on the left side of the river valley a series of river terraces occur. Only a few levels can be followed over longer distances in longitudinal directions. On the right side of the river valley, there are only one or two levels, which can be identified but cannot be traced over longer distances. Ideally, the age assigned to any river terrace is based on some tangible parameter, such as its stratigraphic relation to dated deposits, the radiometric dating of bones of wood included in the terrace sediments, or the degree of weathering of the terrace deposits. Unfortunately, most terrace remnants do not readily end themselves to such straightforward analysis as they contain no datable materials, have complex weathering profiles and, often standing isolated, they defy easy correlation with other features with record established events. In practice, therefore, dating of terraces is often prone to be an educated guess based on elevation above the present stream or some other questionable techniques.

The age of the higher levels is less certain. Further up the Rhine valley, Hüntwangen level, this is comparable with the Münster level (12000 BC), which

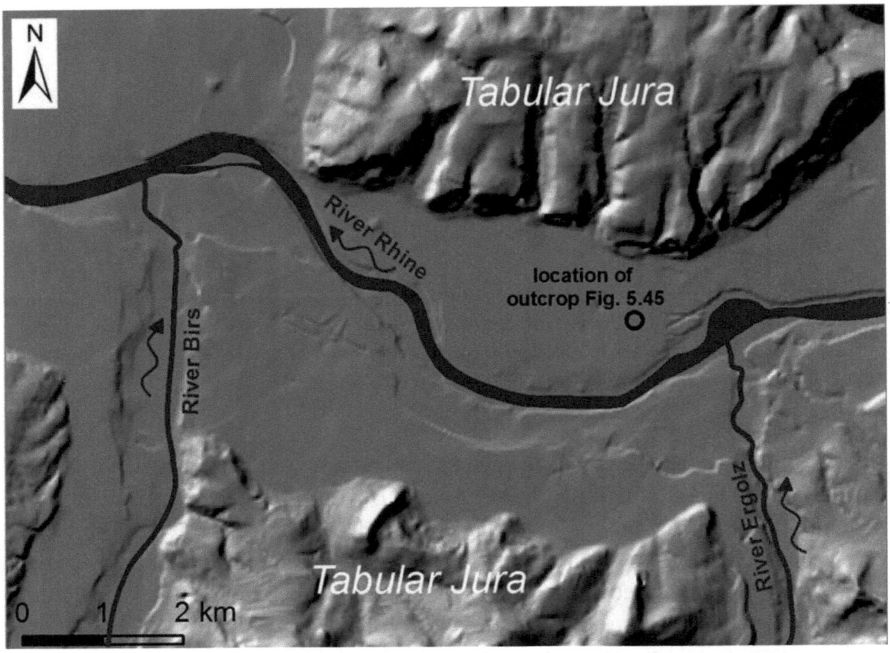

Fig. 5.44 *Shaded display* of the DTM-AV in the area of Basel. The braided river topography is well conserved in the forested area. (D) drainage gully; (S) spring brook, formed by backwards erosion due to exfiltering of groundwater. North is to the top. Reproduced by permission of Swisstopo (BA081589)

age is estimated in the order of 18000 BC (Kock et al. 2009). A decrease in the ages of terrace levels from proximal (the glacier front) to distal (Basel) indicates a combined action of different processes. In fact, tectonic activity may represent one possible reason for base level adjustments, but other processes such as the influence of variable resistance of the unconsolidated rock basis to erosion and the influence of large low frequency floods have be taken into account for large events occurring over time periods of 500 to several 1,000 of years.

There is evidence that river terraces developed during a period of changing geomorphic conditions under the action of lager flood events, which formed a characteristic sequence of river terraces during a rapidly declining base level. Within these river terraces we found some outcrops showing sedimentary features deviating from normal braided river sediments.

Furthermore, we compared the sedimentary structures of outcrops with modern analogs of extraordinary floods (Russell and Knudsen 1999). An impressive outcrop documenting an extreme flood event is located at a distance of 300 m from the right bank of the river Rhine in Grenzach-Whylen, north of Basel (Fig. 5.45). The outcrop consists of a sequence of mixture of trough-cross-bedding (mainly OW-Bimodal gravel) and gray gravel sequences at the base, followed by a distinct erosion surface. On the erosion surface we found blocks and boulders of up to more than 1 m in

Fig. 5.45 Outcrop documenting an extreme flood event located at a distance of 300 m from the right bank of the river Rhine in Grenzach-Whylen, north of Basel (see Fig. 5.44 for location)

diameter consisting of Buntsandstein, Trigonodusdolomites from the Triassic and Upper Jurassic limestones, as well as disturbed gravel "pockets," which internal structure indicate a transport of larger blocks of frozen gravel and subsequent deposition as frozen blocks on the erosional surface. Above these horizons follow a sequence of 2 to 5 m thick moderately sorted, clast-supported gravel with larger clasts and generally with well-developed fabrics. These gravels show slightly horizontal layering but no evidence of erosion surfaces. The gravel sheets of the Hüntwangen event and those of the Grenzach-Whylen outcrops are interpreted as high-density bedload movements (traction carpets) driven by suspension rich stream flow (Todd 1989). As a consequence the whole sequence is likely to represent the result of one single event. This indicates a flood during wintertime or snow melt period. The processes of shearing and cutting blocks of frozen gravel were recently observed during the Skeidararhlaup in 1996 (Smith et al. 2006; Russell and Knudsen 1999). Similar deposits have been previously described by different authors among them Nemec and Muszynski (1982), Hein and Walker (1977), or Reid and Frostick (1989). Stream driven gravely traction carpets produce sheet-like units of clast- to matrix-supported conglomerate, characterized by a parallel fabric. Gravel entrainment, suspension, and traction carped development are significantly easier if the floodwater already carries a high concentration of sand and silt in suspension. According to Todd (1989) gravely traction carpets can be maintained in channels of relatively low gradients by the shear stress exerted by the high-density, sand bearing turbulent flood flow above. This tangential shear stress is converted to dispersive pressure, which aids buoyancy and quasi-static grain-to-grain contacts in the support of clasts within the carpet. The carpet is thought to have a quasi-plastic rheology but behaves much like a viscous fluid at high shear rates.

This process forms a sheet-like homogeneous unit. In comparison to horizontal-bed elements such as diffuse gravel sheets (Siegenthaler and Huggenberger 1993), these elements can laterally extend to hundreds of meters and reach thicknesses up to 5 m.

A comparison of the distribution of river terraces of the River Rhine and the Truckee River demonstrate some structural similarities in the shapes of the terraces.

In the example described by Born and Ritter (1970) terraces developed during a period of changing geometric conditions, a geologic setting different from that normally accepted as controlling the origin of erosional terraces. Evidence presented by Born and Ritter (1970) demonstrates that a complex series of fluvial terraces can develop in an extremely short period of time and, furthermore, that the controls of this development are basically the interplay of extreme floods and base level changes.

5.6.3 Conclusions

Earthquake related damage can be mitigated, i.e., by application of building codes to prevent collapse of buildings, failure of lifelines, and destruction of important infrastructure. Microzonation studies can help to optimize the implementation of building codes. Other hazards such as subrosion can only partly be controlled as it may be natural or due to human activities such as changing groundwater flow regimes at different scales. Although the latter occur more localized, it is difficult to take appropriate measures as the processes of dissolution of evaporitic rocks and transport of dissolved evaporites are not influenced by flow but also by density and viscosity of the fluids (see Sect. 5.4).

The events in consequence of the enhanced geothermal reservoir stimulation in Basel reflect an interesting process to be observed in an urban society confronted with a paradigm change with respect to the use of natural resources such as energy. In particular, the risk perception in the DHM project passed through different phases. In a first phase population, stakeholders and people from the local and federal administration where quite optimistic to setup a production facility in a straightforward way. In the retrospective, the spirit may be referred to as "positive illusions," which were not only shared by the stakeholders but also by most of the scientific community, even though many scientists were aware of the possibility of induced seismicity from published events (Healy et al. 1979; Hisieh and Bredehoeft 1981; Nicolson and Wesson 1992). They all hoped that the hydraulic fracturing does not produce earthquakes that are widely felt or even cause damage, as was then the case. An extensive risk analysis (Baisch et al. 2009), only tasked after the events, showed that also for an area of moderate seismic hazard the risk level is not acceptable for a densely populated area such as the Basel region.

The other important hazard, catastrophic flooding in an urban area of Central Europe can, to some degree, be controlled and regulated through measures at the catchment scale. This includes management of lake levels and temporarily storage in dam lakes or adaption of required space for rivers (i.e., Rhone River). These events represent 100 to 300 year events. However, in the case of the Rhine River, there is evidence from sedimentological and morphological data that a few low frequency events with landscape shaping capacities took place over the last 12,000 years. It is estimated from dated outcrops that the recurrence periods of such events are in the order of 2,000 years, which are comparable to large earthquakes.

References

Ackerer P, Younes A (2008) Efficient approximations for the simulation of density driven flow in porous media. Adv Water Resour 31:15–27. doi:10.1016/j.advwatres.2007.06.001

Affolter A, Huggenberger P, Scheidler S, Epting J (2010) Adaptives Grundwassermanagement in urbanen Gebieten – Ansätze zur konkreten Umsetzung einer nachhaltigen Wasserressourcen-bewirtschaftung. Grundwasser. doi:10.1007/s00767-010-0145-6

Auckenthaler A, Bänninger D, Affolter A, Zechner E, Huggenberger P (2010) Drinking water production close to contaminant sites: a case study from the region of Basel, Switzerland. In: Schirmer M, Hoehn E, Vogt T (eds) Proceedings of the 7th international groundwater quality conference GQ, pp 167–170

Baisch S, Carbon D, Dannwolf U, Delacou B, Devaux M, Dunand F, Jung R, Koller M, Martin C, Sartori M, Secanell R, Vörös R (2009) Deep heat mining basel – seismic risk analysis, Bericht Departement für Wirtschaft, Soziales und Umwelt des Kantons Basel Stadt, Amt für Umwelt und Energie. AP 1000-AP7000

Becker A, Giardini D (2001) Paläoseismologische Forschung in der Schweiz – das PALEOSEIS-Project. GEOForumCH 9:30–33

Becker A, Davenport CA, Haeberli W, Burga C, Perret R, Flisch A, Keller WA (2000) The Fulnau landslide and former Lake Seewen in the northern Swiss Jura Mountains.-. Eclogae Geol Helv 93:291–305

Becker A, Davenport CA, Giardini D (2002) Palaeoseismicity studies on end-Pleistocene and Holocene lake deposits around Basle, Switzerland. Geophys J Int 149:1–20

Bedford B (1996) The need to define hydrologic equivalence at the landscape scale for freshwater wetland mitigation. Ecol Appl 6:57–68

Belchic HC (1960) The Winnfield salt dome, Winn Parish, Louisiana, Guidebook 1960 Spring Field Trip, Shreveport Geological Society, Shreveport , LA, USA, pp 29–48

Blum P, Campillo G, Münch W, Kölbel T (2010) CO_2 savings of ground source heat pump systems – a regional analysis. Renewable Energy 35(1):122–127

Born SM, Ritter DF (1970) Modern terrace development near Pyramid Lake, Nevada, and its geologic implications. GSA Bull 81:1233–1240

Butscher C, Huggenberger P, Auckenthaler A, Bänninger D (2010) Risikoorientierte Bewilligung von Erdwärmesonden (Risk-oriented approval of ground source heat pump systems). Grundwasser. doi:10.1007/s00767-010-0154-5

BUWAL (2004) Wegleitung Grundwasserschutz. Vollzug Umwelt. Bundesamt für Umwelt, Wald und Landschaft, Bern, pp 1–141

Cathomen N, Stauffer F, Kinzelbach W, Osterkorn F (2002) Thermische Grundwassernutzung – Auswirkungen von Wärmepumpanlagen auf die Grundwassertemperatur, gwa 12:901–906

Chiang WH (2005) 3D-Groundwater Modeling with PMWIN: a simulation system for modeling groundwater flow and transport processes, 2nd edn. Springer, Berlin

Costa JE (1987) A comparison of the largest rainfall-runoff floods in the United States with those of the People's Republic of China and the world. J Hydrol 96:101–115

Dijk PE, Berkowitz B (2000) Buoyancy-driven dissolution enhancement in rock fractures. Geology 28(11):1051–1054

Dölemeyer A (2009) Risk and planet earth: an introduction. In: Risk and planet earth, vulnerability, natural hazards, integrated adaption strategies, UNI Leibzig, Schweizerbart, pp 13–17

Dreybrodt W (1990) The role of dissolution kinetics in the development of karstification in limestone: a model simulation of karst evolution. J Geol 98:639–655

Eiswirth M, Hötzl H, Cronin A, Morris B, Veselič M, Bufler R, Burn S, Dillon P (2003) Assessing and improving sustainability of urban water resources and systems. RMZ Mater Geoenviron 50:117–120

Epting J, Huggenberger P, Rauber M (2008a) Integrated methods and scenario development in urban groundwater management, and protection during tunnel road construction; A case study of urban hydrogeology in the city of Basel, Switzerland. Hydrogeol J 16:575–591. doi:10.1007/s10040-007-0242-5

Epting J, Huggenberger P, Regli C, Spoljaric N, Kirchhofer R (2008b) Integrated methods for urban groundwater management considering subsurface heterogeneity. In: Cai X, Jim Yeh TC (eds) Quantitative information fusion for hydrological sciences, vol 79. Springer Series: Studies in Computational Intelligence. Springer, Heidelberg, 218 p. ISBN: 978-3-540-75383-4

Epting J, Huggenberger P, Glur L (2009a) A Concept for Integrated Investigations of Karst phenomena in urban environments – merging geophysical and hydrometrical investigations with 3D hydrogeological modeling for applied urban hydrogeology within a gypsum karst area. Eng Geol. doi:10.1016/j.enggeo.2009.08.013

Epting J, Romanov D, Huggenberger P, Kaufmann G (2009b) Integrating field and numerical modeling methods for applied urban Karst hydrogeology. Hydrol Earth Syst Sci 13:1163–1184. doi:10.5194/hess-13-1163-2009

Epting J, Huggenberger P, Butscher C (2010) Thermal groundwater use in urban areas – Spatio-temporal scales and concepts. In: Schirmer M, Hoehn E, Vogt T (eds) Proceedings of the 7th international groundwater quality conference GQ, pp 67–70

European Environmental Agency EEA (2006) Guidance document for the implementation of the European PRTR. http://www.eea.europe.eu/data-and-maps/esplore-interactive-maps/promotions/european-pollutant-release-and-transfer-register-2014-e-prtr. 148p.

Fäh D (2009) Ein Erdbeben in Augusta Raurica? Jahresberichte aus Augst und Kaiseraugst 30(2009):291–305

Fäh D, Huggenberger P (2006) INTERREG III, Erdbebenmikrozonierung am südlichen Oberrhein. Zusammenfassung für das Projektgebiet Gebiet in der Schweiz. CD and Report (in german; available from the authors)

Fäh D, Wenk T (2009) Mikrozonierung für die Kantone Basel Stadt und Basel Landschaft: Optimierung der Form der Antwortspektren und der Anzahl der Mikrozonen. Abschlussbericht: Teilbericht B Projekt "Umsetzung der Mikrozonierung in den Kantonen Basel Stadt und Basel Landschaft". Schweizerischer Erdbebendienst ETH Zürich

Fäh D, Rüttener E, Noack T, Kruspan P (1997) Microzonation of the city of Basel. J Seismol 1:87–102

Fäh D, Kind F, Lang K, Giardin D (2001) Earthquake scenarios for the city of Basel. Soil Dyn Earthquake Eng 21:405–413

Fäh D, Giardini D, Bay F, Bernardi F, Braunmiller J, Deichmann N, Furrer M, Gantner L, Gisler M, Isenegger D, Jimenez MJ, Kästli P, Koglin R, Masciadri V, Rutz M, Scheidegger C, Schibler R, Schorlemmer D, Schwarz-Zanetti G, Steimen S, Sellami S, Wiemer S, Wössner J (2003) Earthquake catalogue of Switzerland (ECOS) and the related macroseismic database. Eclogae Geol Helv 96:219–236

Fäh D, Steimen S, Oprsal I, Ripperger J, Wössner J, Schatzmann R, Kästli P, Spottke I, Huggenberger P (2006) The earthquake of 250 A.D. in Augusta Raurica, a real event with a 3D site-effect? J Seismol 10:459–477

Fäh D, Gisler M, Jaggi B, Kästli P, Lutz T, Masciadri V, Matt C, Mayer-Rosa D, Rippmann D, Schwarz-Zanetti G, Tauber J, Wenk T (2009a) The 1356 Basel earthquake: an interdisciplinary revision. Geophys J Int 178:351–374

Fäh D, Ripperger J, Stamm G, Kästli P, Burjanek J (2009b) Mikrozonierung für die Kantone Basel Stadt und Basel Landschaft: Validierung und Umsetzung der Mikrozonierung (2006-2008). Abschlussbericht: Teilbericht A Projekt "Umsetzung der Mikrozonierung in den Kantonen Basel Stadt und Basel Landschaft". Schweizerischer Erdbebendienst ETH Zürich

Ferguson G (2009) Unfinished business in geothermal energy. GroundWater 47(2):167

Ferry M, Meghraoui M, Delouis B, Giardini D (2005) Evidence for Holocene paleoseismicity along the Basle-Reinach active normal fault (Switzerland): a seismic source for the 1356 earthquake in the upper Rhine Graben. Geophys J Int 160:554–572

Fujii H, Inatomi T, Itoi R, Uchida Y (2007) Development of suitability maps for ground-coupled heat pump systems using groundwater and heat transport models. Geothermics 36:459–472

Gechter D, Huggenberger P, Ackerer P, Waber HN (2008) Genesis and shape of natural solution cavities within salt deposits. Water Resour Res 44:W11409. doi:10.1029/2007WR006753

Geopower Basel AG (2007) Schadenfälle Erdbeben vom 8.12.2006, Technischer Bericht, Ing-enieurbureau A. Aegerter & Dr. O. Bosshardt AG, Basel, Möhlin, 21p

Gutiérrez F (1996) Gypsum karstification induced subsidence: effects on alluvial systems and derived geohazards (Calatayud Graben, Iberian Range, Spain). Geomorphology 16: 277–293

Hähnlein S, Molina-Giraldo N, Blum P, Bayer P, Grathwohl P (2010) Ausbreitung von Kältefahnen im Grundwasser bei Erdwärmesonden. Grundwasser 15(2):123–133

Harbaugh AW, Banta ER, Hill MC, McDonald MG (2000) MODFLOW-2000, the U.S. geological survey modular ground-water model – user guide to modularization concepts and the ground-water flow process. Open-File Report 00-92. U.S. Geological Survey, Reston, Virginia

Havenith H-B, Fäh D, Polom U, Roulle A (2007) S-wave velocity measurements applied to the seismic microzonation of Basel, Upper Rhine Graben. Geophys J Int 170:346–358

Healy JH, Hamilton RM, Raleigh CB (1979) Earthquakes induced by fluid injection and explosion. Tectonophysics 9:205–214

Hein FJ, Walker RG (1977) Bar evolution and development of stratification in gravelly, braided Kicking Hoarse River, British Columbia. Can J Earth Sci 14:562–570

Hisieh PA, Bredehoeft JD (1981) A reservoir analysis of the Denver earthquakes: a case of induced seismicity. J Geophys Res 86:903–920

Hoehn E (2005) Flussnahes Grundwasser strömt in Etagen. gwa 11: p 905

Huggenberger P, Epting J, Regli C, Spottke I, Zechner E (2006) Fluss-Grundwasser Interaktion. MoNit: Modellierung der Grundwasserbelastung durch Nitrat im Oberrheingraben. INTERREG III A-Projekt. Landesamt für Umwelt, Messungen und Naturschutz Baden-Württemberg, Karlsruhe, 110 Seiten

Huggenberger P, Epting J, Affolter A, Zechner E (2010) Concepts for transboundary groundwater management in a region of extensive groundwater use and numerous contaminated sites. In: Proceedings of the international conference "Transboundary aquifers: challenges and new directions" (ISARM2010)

INTERREG III A-Projekt MoNit (2006) Modellierung der Grundwasserbelastung durch Nitrat im Oberrheingraben. Landesanstalt für Umwelt, Messungen und Naturschutz Baden-Württemberg

Kaltofen R, Opitz R, Schumann K, Ziemann J (1994) Tabellenbuch Chemie. Harri Deutsch, Thun und Frankfurt am Main

Kind F (2002) Development of microzonation methods: application to Basel, Switzerland, Ph.D. thesis, ETH Zürich. Available in electronic form from www.ethbib.ethz.ch

Kind F, Fäh D, Giardini D (2005) Array measurements of s-wave velocities from ambient vibrations. Geophys J Int 160:114–126

Kock S, Huggenberger P, Preusser F, Rentzel P, Wetzel A (2009) Formation and evolution of the Lower Terrace of the Rhine River, in the area of Basel. Swiss J Geosci 102:307–321

Konz M, Ackerer P, Meier E, Huggenberger P, Zechner E, Gechter D (2008) On the measurement of solute concentrations in 2-D flow tank experiments. Hydrol Earth Syst Sci 12:727–738

Konz M, Ackerer P, Younes A, Huggenberger P, Zechner E (2009a) Two-dimensional stable-layered laboratory-scale experiments for testing density-coupled flow models. Water Resour Res 45:W02404. doi:10.1029/2008wr007118

Konz M, Younes A, Ackerer P, Huggenberger P, Zechner E (2009b) Variable-density flow in heterogeneous porous media – Laboratory experiments and numerical simulations. J Cont Hydrol. doi:10.1016/j.jconhyd.2009.07.005 DOI:dx.doi.org

Kozel R (2005) Viele Pumpbrunnen stehen zu nahe am Fluss. Aquaterra 2:11–12

Mandelbrot BB, Wallis JR (1968) Noah, Joseph, and operational hydrology. Water Resour Res 4(5):909–918
Mayer Rosa D, Cadiot B (1979) A review of the 1356 Basel earthquake: basic data. Tectonophysics 53:325–333
Meghraoui M, Delouis B, Ferry M, Giardini D, Huggenberger P, Spottke I, Granet M (2001) Active normal faulting in the Upper Rhine Graben and paleoseismic identification of the 1356 Basel Earthquake. Science 293:2070–2073
Mosher SJ, Martini IP (2002) Coarse-Grained flood bars formed at the confluence of two subarctic rivers affected by hydroelectric dams, Ontario, Canada. Spec Publ Int Ass Sediment 32:213–231
Nakatsuka K (1999) Field characterization for HDR/HWR: a review. Geothermics 28(4–5): 519–531
Nemec W, Muszynski A (1982) Volcaniclastic alluvial aprons in the Tertiary of the Sofia district (Bulgaria). Annales Societatis Geologorum Poloniae 52:239–303
Nicolson C, Wesson RL (1992) In: McGarr A (ed) Triggered earthquakes and deep well activities. Special Issue on induced seismicity. PAGEOPH 139:72–86
OECD (1988) The international hydrocoin project, level 1: Code verification. In: Hydrocoin project. OECD, Paris
Oprsal I, Fäh D (2007) 1D vs 3D strong ground motion hybrid modelling of site, and pronounced topography effects at Augusta Raurica, Switzerland – Earthquakes or battles? In: 4th international conference on earthquake geotechnical engineering, 25–28 June 2007, Paper No. 1416
Oprsal I, Fäh D, Mai M, D Giardini (2005) Deterministic earthquake scenario for the Basel area – Simulating strong motion and site effects for Basel, Switzerland. J Geophys Res 110:04305. doi:10.1029/2004JB003188
Pannike S, Kölling M, Panteleit B, Reichling J, Scheps V, Schulz HD (2006) Auswirkung hydrogeologischer Kenngrößen auf die Kältefahnen von Erdwärmesondenanlagen in Lockersedimenten. Grundwasser 11(1):6–18
Pfister C (1999) Wetternachhersage, 500 Jahre Klimavariationen und Naturkatastrophen, Haupt. 304p
Prokop G (2003) Sustainable management of soil and groundwater resources in urban areas. In: Proceedings of the 2nd IMAGETRAIN cluster meeting, Krakow, Poland, 2–4 Oct 2002
Regli C, Huggenberger P (2006) Entwicklung nachhaltiger Strategien für den Grundwasserschutz. In: Revitalisierung urbaner Flusslandschaften, Schlussbericht zum MGU-Forschungsprojekt F1.03: 47–61
Regli C, Huggenberger P (2007) Grundwasserschutz bei wasserbaulichen Eingriffen in Fliessgewässer – Hydrogeologische Aspekte, gwa 7, 261–272
Regli C, Guldenfels L, Huggenberger P (2004) Revitalisierung von Fliessgewässern im Konflikt mit der Grundwassernutzung, gwa 4, 261–272
Reid I, Frostick LE (1989) Dynamics of bedload transport in Turkey brook, a coarse-grained alluvial channel. Earth Surf Process Landform 11:143–155
Rentzel P (1994) Geologisch-bodenkundliche Untersuchungen an den Niederterrassenfeldern bei Basel unter besonderer Berücksichtigung der spätlatènezeitlichen Fundstelle Basel-Gasfabrik. Jahresbericht Archäologische Bodenforschung des Kanton Basel-Stadt 1994:31–52
Ripperger J, Kästli P, Fäh D, Giardini D (2009) Ground motion and macro-seismic intensities of a seismic event related to geothermal reservoir stimulation below the city of Basel – observations and modelling. Geophys J Int 179:1757–1771
Romanov D, Gabrovsek F, Dreybrodt W (2007) Leakage below dam sites in limestone terrains by enhanced corrosion: a modeling approach. Environ Geol 51:775–779
Russell AJ, Knudsen Ô (1999) Controls on the sedimentology of November 1996 jôkulhlaup deposits, Skeiôarârsandur, Iceland. In: Smith ND, Rogers J (eds) Fluvial sedimentology VI. International Association of Sedimentologists Special Publication No. 28, pp 315–329
Sanner B, Karytsas C, Mendrinos D, Rybach L (2003) Current status of ground source heat pumps and underground thermal energy storage in Europe. Geothermics 32(4–6):579–588

SIA (2003) Norm 261. Einwirkungen auf Tragwerke, Schweizerischer Ingenieur- und Architektenverein (SIA), Zürch, 75 p

Siegenthaler C, Huggenberger P (1993) Pleistocene Rhine gravel: deposits of a braided river system with dominant pool preservation. In: Best JL, Bristow CS (eds) Braided rivers. Geological Society 75:147–162

Smith LC, Sheng Y, Magilligan F, Smith ND, Gomez B, Mertes L, Krabill EB, Garvin JB (2006) Geomorphic impact and rapid subsequent recovery from the 1996 Skeiðarársandur jökulhlaup, Iceland, measured with multi-year airborne lidar. Geomorphology 75:65–75

Spottke I, Zechner E, Huggenberger P (2005) The southeastern border of the Upper Rhine Graben: a 3D geological model and its importance for tectonics and groundwater flow. Int J Earth Sci 94:580–593. doi:10.1007/s00531-005-0501-4

Todd SP (1989) Stream-driven, high-density gravelly traction carpets: possible deposits in the Trabeg Conglomerate formation, SW Ireland and some theoretical considerations of their origin. Sedimentology 36:513–530

United Nations for Environment Programme/Agency for the Environment and Energy Management (2005) Identification and management of contaminated sites – a methodological guide. UNEP, Washington DC, 132 p

Wagner U, Huggenberger P, Schaub D, Thater M (2001) Interreg III: Erkundung der Grundwasserleiter und Böden im Hochrheintal [Interreg III: investigation of the aquifers and soils in the Hochrheintal]. LandratsamtWaldshut, Waldshut, Germany, 101 pp

Watson SJ, Barry DA, Schotting RJ, Hassanizadeh SM (2002) Validation of classical density-dependent solute transport theory for stable, high-concentration-gradient brine displacements in coarse and medium sands. Adv Water Resour 25:611–635

Wells MJM, Fono LJ, Pellegrin ML (2007) Water Environ Res 79(10):2192–2209

Wiemer S, Giardini D, Fäh D, Deichmann N, Sellami S (2008) Probabilistic seismic hazard assessment of Switzerland: best estimates and uncertainties. J Seismol. doi:10.1007/s10950-008-9138-7

Wittmann O (1961) Die Niederterrassenfelder im Umkreis von Basel und ihre kartographische Darstellung. Basler Beiträge zur Geographie und Ethnologie 3:46

Zechner E, Kind F, Fäh D, Huggenberger P (2001) 3-D Geological model of the Southern Rhine Graben compiled on existing geological data and geophysical reference modeling. In: Abstract Volume of the 2nd EUCOR-URGENT Workshop, 7–11 Oct 2001, Mont Saint-Odile, Strasbourg, France, p 43

Glossary

Adaptive Water Management (AWM) Adaptive management refers to a systematic process for continually improving management policies and practices by learning from the outcomes of implemented management strategies (Pahl-Wostl 2007).

Alluvial valley A valley filled with the water-borne sediments composed of unconsolidated alluvial material.

Aquiclude An aquiclude is a solid, impermeable area underlying or overlying by an aquifer.

Aquifer A relatively highly permeable water-bearing geological formation underlain by a less permeable layer (aquitard or aquiclude) and the water contained in the saturated zone of the formation (UN Doc. A/RES/51/229).

Aquifer storage capacity Aquifer storage capacity is defined as the maximum volume of water that can be stored in an aquifer (Loáiciga 2008).

Aquifer system Means a series of two or more aquifers that are hydraulically connected (UN Doc. A/RES/51/229).

Aquitard An aquitard is a relatively poorly permeable water-bearing geological formation capable of retaining water of an overlain aquifer.

Artesian groundwater Artesian groundwater derives from a confined aquifer containing water under positive pressure. In case of tapping, this causes the water level to rise to the surface. In case the water level does not reach the surface, the term confined groundwater is used.

Artificial groundwater recharge Artificial groundwater recharge comprises the reinjection of, e.g., surface water into an aquifer. Artificial recharge can contribute substantially to the overall groundwater budget; methods include (1) injection wells and infiltration ditches; (2) artificial infiltration ponds with filtering layers; and (3) infiltration fields.

Attenuation The process by which a compound is reduced in concentration over time, through adsorption, degradation, dilution, and/or transformation.

Baseflow Baseflow is the portion of streamflow coming from the longer-term discharge into a stream from natural storages, notably sustaining flow between rainfall events. Certain parameters of baseflow, such as the mean residence time

P. Huggenberger and J. Epting (eds.), *Urban Geology: Process-Oriented Concepts for Adaptive and Integrated Resource Management*, DOI 10.1007/978-3-0348-0185-0,

and the baseflow recession curve, can be useful in describing the mixing of waters (such as from precipitation and groundwater) and the level of groundwater contribution to streamflow in catchments (Vitvar et al. 2002).

Bedload movements Bedload is the sand, gravel, boulders, or other debris transported by rolling or sliding along the bottom of a stream. Bedload moves by rolling, sliding, and/or saltating.

Benchmark experiments A problem designed to evaluate the performance of a computer system or of different candidate algorithms and the identification of the best among them. The comparison of algorithms with respect to point estimates of performance measures, for example, computed via cross-validation, is an established procedure in benchmark studies.

Body of groundwater (GWB) The management of resources in urban areas implies a definition of manageable units of the subsurface. The Water Framework Directive (WFD) introduced a notion of a body of groundwater (GWB) with a new definition – "a GWB is a distinct volume of groundwater within an aquifer or aquifers". According to the WFD, a GWB will be a management unit of groundwater necessary for a subdivision of aquifers in order for them to be effectively managed.

Boundary conditions The set of conditions specified for the behavior of the solution to a set of differential equations at the boundary of its domain. Boundary conditions are important in determining the mathematical solutions to many physical problems. For numerical simulation a region of interest is delineated which has a certain boundary with the surrounding environment. Here the physical processes in the boundary region (head, flux) can be defined.

Breakthrough events Moments when at the outflow of karst systems a sudden change in discharge can be observed. Breakthrough events come along with the evolution of the karst system and events when one or more conduits break through.

Breakthrough time Point in time at which a change in concentration or hydraulic head or discharge induced at a point of origin in the flow domain arrives, i.e., becomes observable; at a point of observation (see also breakthrough events).

Brine Saline water with total dissolved salt content exceeding that of seawater, i.e., approximately 35,000 mg/l (Toth 2009). For example, NaCl-saturated brine from solution-mined salt caverns contains: NaCl = 300–310 g/l, $CaSO_4$ = 5.0–5.5 g/l, $MgSO_4$ = 0.3–0.4 g/l, and $MgCl_2$ = 0.1–0.2 g/l (specification of NaCl-saturated brine; code-No. 9400; United Swiss Saltworks).

Capture zone A capture zone is the aquifer volume through which groundwater flows to a pumping well over a given time of travel. Determining a well's capture zone aids in water supply management by creating an awareness of the water source. This helps ensure sustainable pumping operations and outlines areas where protection from contamination is critical (Ahern et al. 2003).

Channel belt Part of alpine and perialpine type of valley fills, consisting of sediments transported or deposited in the channels at different flow stages (mainly gravel deposits).

Climate change Climate change refers to a change in the state of the climate that can be identified (e.g., by using statistical tests) by changes in the mean and/or the variability of its properties and that persists for an extended period, typically decades or longer.

Correlation lengths The correlation function is a measure of the order in a system, as characterized by a mathematical correlation function, and describes how, e.g., sedimentary structure types at different positions are correlated. Spatial correlation lengths allow to characterize the geometric anisotropy of sedimentary structure types.

Data mining Efficient data management that allows fast access to different types of, in our case, geological and hydrogeological data for further data analysis, setup of geological or hydrogeological models, and the definition of boundary conditions.

Deep-seated evaporites Evaporitic rocks such as rock salt (halites), Anhydrite, or Gypsum that do not occur near the surface.

Drains A drain is a natural or artificial system that dewaters (drains) surface areas or aquifer volumes in the subsurface.

Dual (mixed) flow systems Diffuse (slow) and conduit (fast) components groundwater flow in karst systems.

Dynamic Vulnerability Index DVI DVI is defined as the ratio of the contributions of conduit and diffuse systems to spring discharge. It serves as a quantitative indicator of the intrinsic vulnerability of a karst spring (Butscher and Huggenberger 2008).

Earthquake magnitudes Earthquake magnitude is a measure of earthquake size and is determined from the logarithm of the maximum displacement or amplitude of the earthquake signal as seen on the seismogram, with a correction for the distance between the focus and the seismometer. This is necessary as the closer the seismometer is to the earthquake, the larger the amplitude on the seismogram, irrespective of the size or magnitude of the event. Since the measurement can be made from P, S, or surface waves, several different scales exist, all of which are logarithmic because of the large range of earthquake energies (for example, a magnitude 6 ML is 30 times larger, in terms of energy than a magnitude 5 ML). The Richter local magnitude (ML) is defined to be used for "local" earthquakes up to 600 km away. Moment magnitude (Mw) is considered the best scale to use for larger earthquakes. Moment magnitude is measured over the broad range of frequencies present in the earthquake wave spectrum rather than the single frequency sample that the other magnitude scales use. http://www.earthquakes.bgs.ac.uk/earthquakes/education/faqs/faq15.html

Earthquake microzonation In most general terms, seismic microzonation is the process of estimating the response of soil layers under earthquake excitations and thus the variation of earthquake characteristics on the ground surface. Microzonation provides the basis for site-specific risk analysis, which can assist in the mitigation of earthquake damages.

Emerging contaminants Defined as compounds that are not currently covered by existing regulations of water quality, that have not been previously studied, and that are thought to be a possible threat to environmental health and safety (Ferrer and Thurman 2003).

Environmental management scheme A mechanism by which landowners and other individuals and bodies responsible for land and natural resources management can be incentivized to manage their land in a manner sympathetic to the environment.

Endogenic processes All processes that take place inside Earth (and other planets) are considered endogenous. They make the continents migrate, push the mountains up, and trigger earthquakes and volcanism. Endogenous processes are driven by the warmth that is produced in the core of Earth by radioactivity and gravity.

Equivalence and acceptance criteria Equivalence and acceptance criteria (Bedford 1996) allow comparing different development scenarios. They can assess the technical benefits of different engineering projects, the supply situation for industrial groundwater users, the development of the groundwater flow regime, and the improvement of overall groundwater quality.

Eurocode 8: Earthquake resistance The standards of Eurocode 8 apply to the design of buildings and civil engineering works in seismic regions with the aim of ensuring that human lives are protected, damage is limited, and structures important for civil protection remain operational. (http://www.eurocodes.co.uk/EurocodeDetail.aspx?Eurocode=8)

Exfiltration Movement of water or mass from the hyporheic zone (see below) and groundwater to surface waters.

Exogenic processes Geology formed or occurring on the surface of the earth (e.g., flooding and erosion).

Extensometer measurements An extensometer is a device that is used to measure changes in the length of an object. It is useful for natural hazard measurements including, e.g., subsidence and landslides (see also inclinometer).

Flood protection Precautionary measures, equipment, or structures implemented to guard or defend people, property, and lands from an unusual accumulation of water above the ground.

Flow across boundaries Heat and mass fluxes including water compounds across the boundaries of groundwater bodies (see GWB).

Flux The rate of flow, energy, or water compounds across a given surface.

Geodatabase A database designed for data mining (see above) including the storage and query of different types of geological and hydrogeological information.

Geothermal energy Refers to energy stored in the form of heat in solid formations beneath the earth's surface, which has primarily been formed from the decay of naturally occurring radioactive elements. The temperature is more or less constant all year round from a depth of approximately 15 m below the surface.

Goals for sustainable development Goals include the sustainable use of subsurface resources and take into account long-term impact of geotechnical measures and future changes in usage. Whereas goals focus on a long-term sustainable development, milestones (see below) center on short-term subsurface resource protection and geotechnical issues.

Groundwater flow regime The terminology includes groundwater flow patterns, velocities, and budgets for a defined region in a temporal context (Epting et al. 2007).

HACCP and CCPs The setup of management systems based on the Hazard Analysis and Critical Control Points concept (HACCP) incorporate strategies to deal with day-to-day management of water quality, including upsets and failures (Rauch 2009). Critical Control Points (CCPs) mark moments of relevant levels concerning the quality of drinking water, where it is possible and necessary to avoid a health risk and to eliminate the risk or minimize the impacts to an acceptable level. CCPs allow to identify the risk of pollution and to adapt the control and operation system (e.g., depending on river–groundwater interaction, e.g., Rauch 2009).

Hard and soft data The distinction of hard and soft data allows categorizing the reliability of information derived from data. Hard data can be derived, e.g., from outcrop and laboratory investigations, which generally can be repeated. Soft data as derived, e.g., from borehole descriptions and geophysical data, provide limited information on the spatial distribution of subsurface properties and permit speculative conclusions.

Hazards/Geohazards A hazard is a situation that poses a level of threat to life, health, property, or environment. A geohazard can be defined as a geological state that represents or has the potential to develop further into a situation leading to damage or uncontrolled risk.

Headwaters The tributary streams of a river in the area in which it rises; headstreams.

Hydraulic base failure Hydraulic base failure can occur when hydrostatic uplift affects the stability of subsurface constructions. In this case, springs can break up directly in the foundation of subsurface constructions.

Inclinometer An inclinometer is an instrument for measuring angles of slope (or tilt), elevation, or depression of an object with respect to gravity. It is useful for natural hazard measurements including, e.g., subsidence and landslides (see also extensometer).

Indicator substances Substances that indicate the presence, absence, or concentration of another substance or the degree of reaction between two or more substances by means of a characteristic change.

Infiltration Movement of water or mass from surface waters to the hyporheic zone (see above) and groundwater.

Interstitial (hyporheic zone) The interstitial or hyporheic zone is the pore-space in the riverbed, under or beside a stream channel, or floodplain that contributes water to the stream. Hyporheic flow, also called interstitial flow, is the subsurface flow between the water table and surface water flow. The source of hyporheic flow can be from the channel itself or the water percolating to the stream from the surroundings. The water flow within the hyporheic zone can be considerably high and even as large as in the stream itself. Hyporheic flow may comprise all of the flow in arid areas with sandy soils, such as in desert areas, when the surface waters already have dried up.

Intrastratal Between stratigraphic units.

Intrinsic and specific vulnerability Vrba and Zoporozec (1994) defined "intrinsic vulnerability" as an intrinsic property of a groundwater system and one that depends on the sensitivity of that system to human or natural impacts, whereas "specific vulnerability" was defined as the risk of pollution due to the potential impact of specific land uses and contaminants (Martínez-Bastida et al. 2010).

Laminar to turbulent Laminar or streamline flow occurs when a fluid flows in parallel layers, with no disruption between the layers. Turbulent flow is a flow regime characterized by chaotic and stochastic property changes. This includes low momentum diffusion, high momentum convection, and rapid variation of pressure and velocity in space and time.

Leakage Movement of water or mass from one system to another, e.g., surface water to groundwater or vice versa and between aquifers.

Liquefaction Liquefaction of saturated, unconsolidated sediments is the process by which very wet sediment starts to behave like a liquid. Liquefaction occurs because of the increased pore pressure and reduced effective stress between solid particles generated by the presence of liquid. It is often associated by severe shaking as caused by earthquakes.

Milestones Milestones represent moments when available knowledge is evaluated with respect to decisions and based on previously defined criteria. This includes the minimization of qualitative or quantitative changes of surface waters and subsurface resources, safeguarding water quality measures during water engineering projects, as well as the development of technical solutions that guarantee sustainable development of the subsurface.

Multivariate statistics Multivariate statistics is a form of statistics encompassing the simultaneous observation and analysis of more than one statistical variable. Methods in urban hydrogeology can include the application of principal component analysis (PCA) or a combination of self-organizing maps (SOM) with Sammon's Projection (SOM-SM).

Open sump drainage Open sump drainage is a pumping technique applied to dewater construction sites that are located beneath the groundwater table without further geotechnical measures. This technique induces a far wider-ranging drawdown compared to dewatering techniques where residual groundwater is pumped from areas enclosed, e.g., by sheet pile walls.

Paleochannels Deposits of unconsolidated sediments or semiconsolidated sedimentary rocks deposited in ancient, currently inactive river and stream channel systems.

Percolating precipitation water Groundwater is replenished or recharged by percolating precipitation water, rainwater, and melting snow that soak into the soil. This water percolates downward and eventually reaches the water table.

Protective cover layers Groundwater protective layers are defined as a naturally occurring soil or geologic layer or series of layers capable of significantly impeding downward movement of water and contaminants (Pest et al. 1993). This can be accomplished by limiting the rate of movement or by degradation, adsorption, or retardation of chemical and microbiological compounds.

Proxy A proxy variable is something that is probably not in itself of any great interest, but from which a variable of interest can be obtained. In order for this to be the case, the proxy variable must have a close correlation, not necessarily linear or positive, with the inferred value.

Quality control systems The present approach of the WHO drinking water guidelines to drinking water supply management includes a systematic assessment of risks throughout a drinking water supply – from the catchment and its source water through to the consumer – and identifying the ways in which these risks can be managed. This also includes methods to ensure that control measures are working effectively. Quality control systems include the monitoring of physical, chemical, and microbiological parameters, the definition of Critical Control Points (CCPs, see above), as well as flux calculations, which can be derived from groundwater modeling.

Residence time Residence time is the average time a chemical or microbiological compound of interest spends in a groundwater or surface water system.

Riparian zone A riparian zone or riparian area is the interface or corridor between land ecosystems and riverbed environments. Riparian zones are significant in ecology, environmental management, and civil engineering because of their role in soil conservation, their habitat biodiversity, and the influence they have on fauna and aquatic ecosystems.

Risk Risk is defined as the product of the probability or frequency of an event to happen times the magnitude of damage.

Risk assessment Risk assessment is a step in a risk management procedure. Risk assessment is the determination of quantitative or qualitative value of risk related to a concrete situation and a recognized threat (also called hazard).

Risk profiles The concept of "risk profiles" includes the determination of principal hazards and related damage scenarios for subsurface resources or different water supplies. This includes quantitative and qualitative degradations of urban subsurface resources and the potential impact or damage for water supplies and consumers.

River terrace Level land terrace formed in a valley by fluviatile erosion, aggradation, or tectonic activities (Pfannkuch 1990).

River/stream restoration River restoration describes a set of activities that help improve the environmental health of a river or stream. Improved health may be indicated by expanded habitat for diverse species (e.g., fish, aquatic insects, and other wildlife) and reduced stream bank erosion (MCDEP 2010). Enhancements may also include improved water quality (i.e., reduction of pollutant levels and increased dissolved oxygen levels) and achieving a self-sustaining, functional flow regime in the stream system that does not require periodic human intervention, such as dredging or construction of flood control structures (Gilman and Jarrod 2009).

Riverbed clogging and conductance Clogging of a riverbed refers generally to changes in the exchange processes between the river water and groundwater. These processes are usually described as infiltration and filtering, and they are accompanied by changes in: flow throughout the riverbed, mechanical filtering or sieving, sorption, chemical oxidation and reduction, and ion exchange (Mucha et al. 2006). The conductance is the resulting permeability or resistance for in- or exfiltrating processes (see above) of the riverbed.

Rock swelling Rock swelling occurs during the transformation of anhydrite into gypsum by the uptake of water, which is accompanied by a considerable rock volume increase associated with an increase of the swelling pressure which can lead to terrain uplift.

Scenario development Scenarios represent system states and event sequences and serve to acquire and illustrate a representative selection of possible dispositions and process sequences. Scenario development also involves the simplification and restriction of essential boundary conditions that affect the system. With calibrated models scenarios can be developed which facilitate the evaluation of system sensitivities, allowing the investigation of certain parameters and boundary conditions.

Sealing and sheet pile wall Technical measures to stabilize the subsurface and to avoid or minimize groundwater inflow during construction measures.

Source and resource protection Vulnerability assessment methods (see below) can be divided into "source" and "resource protection" methods (Hötzl 1996). Resource protection methods aim to protect all of the groundwater, whereas source protection methods focus on the protection of a discrete water source.

Subrosion Subsurface salt (halite) dissolution or "subrosion" (term used in German literature since 1926; Gechter 2008).

Subsidence Subsidence is the motion of a surface (usually, the Earth's surface) as it shifts downward relative to a datum such as sea-level. Subsidence frequently causes hazards in karst terrains, where dissolution of carbonates or evaporites by fluid flow in the subsurface causes the creation of voids (i.e., caves).

System analysis System analysis includes the documentation of the current system state or profile (see below) as well as stationary or nonstationary processes taking place in the system.

System profiles System profiles describe the state of an environmental system at a particular moment.

Thermal regime The terminology includes the distribution of thermal patterns and budgets, including advective and conductive heat transport, for a defined region in a temporal context.

Transient hydrogeology In transient hydrogeological environments not only the natural and anthropogenic boundary conditions are spatiotemporal variable but also the aquifers and aquifer properties in general, e.g., in systems were karst develops.

Trough-cross-bedding Sedimentary structure of coarse fluvial gravel deposits consisting of large cross-bedded sets with trough-shaped, concave upward, and erosive lower bounding surfaces.

Urban geology Urban geology is a branch of geology that provides information required for sound urban planning and sustainable development in densely populated areas. It includes the participation of diverse branches of earth sciences such as geology, hydrology, or engineering, which allow understanding of the geological and hydrogeological processes in dynamic urban environments.

Use conflicts Conflicts arising from different interests with respect to planned activities including effects on environmental issues.

Vulnerability assessment methods Many methods have been developed to obtain contamination vulnerability assessments of groundwater, most of these being based on overlay and index techniques, and relying on the quantitative or semiquantitative compilation and interpretation of mapped data (Gogu and Dassargues 2000).

Vulnerability mapping A practical tool to implement strategies for land-use planning and sustainable socioeconomic development coherent with groundwater protection. The objective of vulnerability mapping is to identify the most vulnerable zones of catchment areas and to provide criteria for protecting the groundwater used for drinking water supply (Marín et al. 2010).

Water engineering measures Measures that include flood control, construction site drainages, construction parts reaching into the aquifer, and storm water management.

Water Framework Directive (WFD) Directive 2000/60/EC of the European Parliament and of the Council from 23 October 2000 establishing a framework for Community action in the field of water policy.

References

Ahern JA, Lilly MR, Hinzman LD (2003) Ground-water capture zone delineation of hypothetical systems: methodology comparison and real-world applications. American Geophysical Union, Fall Meeting 2003, abstract #H21E-0892

Bedford B (1996) The need to define hydrologic equivalence at the landscape scale for freshwater wetland mitigation. Ecol Appl 6:57–68

Butscher C, Huggenberger P (2008) Intrinsic vulnerability assessment in karst areas: a numerical modeling approach. Water Resour Res 44:W03408

Epting J, Regli C, Huggenberger P (2007) Groundwater protection in urban areas incorporating adaptive groundwater monitoring and management. In: Pahl-Wostl C, Kabat P, Möltgen J (eds) Adaptive and integrated water management, coping with complexity and uncertainty, vol XIV. Springer, Heidelberg, 440 p. ISBN: 978-3-540-75940-9

Ferrer I, Thurman EM (2003) Liquid chromatography/time-of-flight/mass spectrometry (LC/TOF/MS) for the analysis of emerging contaminants. Trends Anal Chem 22(10):2003

Gechter D (2008) Genesis and shapes of salt and gypsum solution cavities created by density-driven groundwater flow: a laboratory experimental approach. Doctoral Thesis at the University of Basel, Basel, Switzerland

Gilman JB, Jarrod K (2009) Challenges of stream restoration as a stormwater management tool. Stormwater 10(3). ISSN: 1531–0574

Gogu RC, Dassargues A (2000) Current trends and future challenges in groundwater vulnerability assessment using overlay and index methods. Environ Geol 39(6):549–559

Hötzl H (1996) Grundwasserschutz in Karstgebieten. Grundwasser 1/96(1996):pp. 5–11

Loáiciga HA (2008) Aquifer storage capacity and maximum annual yield from long-term aquifer fluxes. Hydrogeol J 16:399–403

Marín AI, Andreo B, Mudarra M (2010) Importance of evaluating karst features in contamination vulnerability and groundwater protection assessment of carbonate aquifers. The case study of Alta Cadena (Southern Spain). Zeitschrift für Geomorphologie 54(Suppl. 2):179–194

Martínez-Bastida JJ, Arauzo M, Valladolid M (2010) Intrinsic and specific vulnerability of groundwater in central Spain: the risk of nitrate pollution. Hydrogeol J 18:681–698

Montgomery County Department of Environmental Protection (MCDEP), Rockville, MD (2010) Benefits of Stream Restoration

Mucha I, Bansky L, Hlavaty Z, Rodak D (2006) Impact of riverbed clogging-colmation- on groundwater. Riverbed Filtration Hydrol 43–72, http://www.springerlink.com/content/g75x883r83j830x7/

Pahl-Wostl C (2007) Transitions towards adaptive management of water facing climate and global change. Water Resour Manag 21:49–62

Pesti G, Bogardi I, Kelly WE, Kalinski RJ (1993) Cokriging of geoelectric and well data to define aquifer protective layers. Ground Water 31(6):905–912

Pfannkuch HO (1990) Elsevier's dictionary of environmental hydrogeology. Elsevier Science, Amsterdam

Rauch W (2009) Anwendung des HACCP Konzepts (Hazard Analysis and Critical Control Points) zum Schutz eines Trinkwasserbrunnens. GWF 07–08

Toth J (2009) Gravitational systems of groundwater flow theory, evaluation, utilization. Cambridge University Press, Cambridge

Vitvar T, Burns DA, Lawrence GB, McDonnell JJ, Wolock DM (2002) Estimation of baseflow residence times in watersheds from the runoff hydrograph recession: method and application in the Neversink watershed, Catskill Mountains, New York. Hydrol Process 16:1871–1877

Vrba J, Zoporozec A (1994) Guidebook on mapping groundwater vulnerability. In: Vrba J, Zoporozec A (eds) International contributions to hydrogeology (IAH), vol 16. IAH, Hannover, p 131

Index

P. Huggenberger and J. Epting (eds.), *Urban Geology: Process-Oriented Concepts for Adaptive and Integrated Resource Management*, DOI 10.1007/978-3-0348-0185-0,

G

H

MIX
Papier aus verantwortungsvollen Quellen
Paper from responsible sources
FSC® C105338

If you have any concerns about our products,
you can contact us on
ProductSafety@springernature.com

In case Publisher is established outside the EU,
the EU authorized representative is:
Springer Nature Customer Service Center GmbH
Europaplatz 3, 69115 Heidelberg, Germany

Printed by Libri Plureos GmbH
in Hamburg, Germany